T0329577

INTRODUCTION TO
RF PROPAGATION

INTRODUCTION TO RF PROPAGATION

John S. Seybold, Ph.D.

WILEY-INTERSCIENCE

JOHN WILEY & SONS, INC.

Published by John Wiley & Sons, Inc., Hoboken, New Jersey
Published simultaneously in Canada

For general information on our other products and services or for technical support, please
contact our Customer Care Department within the United States at (800) 762-2974, outside
the United States at (317) 572-3993 or fax (317) 572-4002.

Wiley also publishes its books in a variety of electronic formats. Some content that appears in
print may not be available in electronic formats. For more information about Wiley products,
visit our web site at www.wiley.com.

Library of Congress Cataloging-in-Publication Data:
Seybold, John S., 1958–
    Introduction to RF propagation / by John S. Seybold.
        p.   cm.
    Includes bibliographical references and index.
    ISBN-13 978-0-471-65596-1 (cloth)
    ISBN-10 0-471-65596-1 (cloth)
    1. Radio wave propagation—Textbooks. 2. Radio wave propagation—Mathematical
models—Textbooks. 3. Antennas (Electronics)—Textbooks. I. Title.
QC676.7.T7S49 2005
621.384'11—dc22

                                                                        2005041617

10 9 8 7 6 5 4 3 2

To:
My mother, Joan Philippi Molitor
and my father, Lawrence Don Seybold

▰▰▰▰ CONTENTS

With the rapid expansion of wireless consumer products, there has been a considerable increase in the need for radio-frequency (RF) planning, link planning, and propagation modeling. A network designer with no RF background may find himself/herself designing a wireless network. A wide array of RF planning software packages can provide some support, but there is no substitute for a fundamental understanding of the propagation process and the limitations of the models employed. Blind use of computer-aided design (CAD) programs with no understanding of the physical fundamentals underlying the process can be a recipe for disaster. Having witnessed the results of this approach, I hope to spare others this frustration.

A recent trend in electrical engineering programs is to push RF, network, and communication system design into the undergraduate electrical engineering curriculum. While important for preparing new graduates for industry, it can be particularly challenging, because most undergraduates do not have the breadth of background needed for a thorough treatment of each of these subjects. It is hoped that this text will provide sufficient background for students in these areas so that they can claim an understanding of the fundamentals as well as being conversant in relevant modeling techniques. In addition, I hope that the explanations herein will whet the student's appetite for further study in the many facets of wireless communications.

This book was written with the intent of serving as a text for a senior-level or first-year graduate course in RF propagation for electrical engineers. I believe that it is also suitable as both a tutorial and a reference for practicing engineers as well as other competent technical professionals with a need for an enhanced understanding of wireless systems. This book grew out of a graduate course in RF propagation that I developed in 2001. The detailed explanations and examples should make it well-suited as a textbook. While there are many excellent texts on RF propagation, many of them are specifically geared to cellular telephone systems and thus restrictive in their scope. The applications of wireless range far beyond the mobile telecommunications industry, however, and for that reason I believe that there is a need for a comprehensive text. At the other end of the spectrum are the specialized books that delve into the physics of the various phenomena and the nuances of various modeling techniques. Such works are of little help to the uninitiated reader requiring a practical understanding or the student who is encountering RF propagation for the first time. The purpose of this text is to serve as a first introduction to RF propagation and the associated modeling. It has been written from the perspective of a seasoned radar systems engineer who sees RF propagation as one of the key elements in system design rather than an end in itself. No attempt has been made to cover all of the theoretical aspects of RF propagation or to provide a comprehensive survey of the available models. Instead my goal is to provide the reader with a basic understanding of the concepts involved in the propagation of electromagnetic waves and exposure to some of the commonly used modeling techniques.

There are a variety of different phenomena that govern the propagation of electromagnetic waves. This text does not provide a detailed analysis of all of the physics involved in each of these phenomena, but should provide a solid understanding of the fundamentals, along with proven modeling techniques. In those cases where the physics is readily apparent or relative to the actual formulation of the model, it is presented. The overall intent of this text is to serve as a first course in RF propagation and provide adequate references for the interested reader to delve into areas of particular relevance to his/her needs.

The field of RF propagation modeling is extremely diverse and has many facets, both technical and philosophical. The models presented herein are those that I perceive as the most commonly used and/or widely accepted. They are not necessarily universally accepted and may not be the best choice for a particular application. Ultimately, the decision as to which model to use rests with the system analyst. Hopefully the reader will find that this book provides sufficient understanding to make the required judgments for most applications.

ACKNOWLEDGMENTS

The most difficult aspect of this project has been declaring it finished. It seems that each reading of the manuscript reveals opportunities for editorial improvement, addition of more material, or refinement in the technical presentation. This is an inevitable part of writing. Every effort has been made to correct any typographical or technical errors in this volume. Inevitably some will be missed, for which I apologize. I hope that this book is found sufficiently useful to warrant multiple printings and possibly a second edition. To that end, I would appreciate hearing from any readers who uncover errors in the manuscript, or who may have suggestions for additional topics.

I have had the privilege of working with many fine engineers in my career, some of whom graciously volunteered to review the various chapters of this book prior to publication. I want to thank my friends and colleagues who reviewed portions of the manuscript, particularly Jerry Brand and Jon McNeilly, each of whom reviewed large parts of the book and made many valuable suggestions for improvement. In addition, Harry Barksdale, Phil DiPiazza, Francis Parsche, Parveen Wahid, John Roach III, and Robert Heise each reviewed one or more chapters and lent their expertise to improving those chapters.

I also want to thank my publisher, who has been extremely patient in walking me through the process and who graciously provided me with two deadline extensions, the second of which to accommodate the impact of our back-to-back hurricanes on the east coast of Florida.

Finally, I want to thank my wife Susan and our children Victoria and Nathan, who had to share me on many weekends and evenings as this project progressed. I am deeply indebted to them for their patience and understanding.

The Mathcad files used to generate some of the book's figures can be found at ftp://ftp.wiley.com/public/sci_tech_med/rf_propagation. These files include the ITU atmospheric attenuation model, polarization loss factor as a function of axial ratio, and some common foliage attenuation models.

JOHN S. SEYBOLD

CHAPTER 1

Introduction

As wireless systems become more ubiquitous, an understanding of radio-frequency (RF) propagation for the purpose of RF planning becomes increasingly important. Most wireless systems must propagate signals through nonideal environments. Thus it is valuable to be able to provide meaningful characterization of the environmental effects on the signal propagation. Since such environments typically include far too many unknown variables for a deterministic analysis, it is often necessary to use statistical methods for modeling the channel. Such models include computation of a mean or median path loss and then a probabilistic model of the additional attenuation that is likely to occur. What is meant by "likely to occur" varies based on application, and in many instances an *availability* figure is actually specified.

While the basics of free-space propagation are consistent for all frequencies, the nuances of real-world channels often show considerable sensitivity to frequency. The concerns and models for propagation will therefore be heavily dependent upon the frequency in question. For the purpose of this text, RF is any electromagnetic wave with a frequency between 1 MHz and 300 GHz. Common industry definitions have RF ranging from 1 MHz to about 1 GHz, while the range from 1 to about 30 GHz is called microwaves and 30–300 GHz is the millimeter-wave (MMW) region. This book covers the HF through EHF bands, so a more appropriate title might have been *Introduction to Electromagnetic Wave Propagation*, but it was felt that the current title would best convey the content to the majority of potential readers.

1.1 FREQUENCY DESIGNATIONS

The electromagnetic spectrum is loosely divided into regions as shown in Table 1.1 [1]. During World War II, letters were used to designate various frequency bands, particularly those used for radar. These designations were classified at the time, but have found their way into mainstream use. The band identifiers may be used to refer to a nominal frequency range or specific frequency ranges

---

Introduction to RF Propagation, by John S. Seybold

TABLE 1.1 Frequency Band Designations

Band	Designation	Frequency Range
Extremely low frequency	ELF	<3 kHz
Very low frequency	VLF	3–30 kHz
Low frequency	LF	30–300 kHz
Medium frequency	MF	300 kHz–3 MHz
High frequency	HF	3–30 MHz
Very high frequency	VHF	30–300 MHz
Ultra-high frequency	UHF	300 MHz–3 GHz
Super-high frequency	SHF	3–30 GHz
Extra-high frequency	EHF	30–300 GHz

TABLE 1.2 Frequency Band Designations

Label	Nominal Frequency Range	ITU—Region 2
HF	3–30 MHz	
VHF	30–300 MHz	138–145, 216–225 MHz
UHF	300–1000 MHz	420–450, 890–942 MHz
L	1–2 GHz	1215–1400 MHz
S	2–4 GHz	2.3–2.5, 2.7–3.7 GHz
C	4–8 GHz	5.25–5.925 GHz
X	8–12 GHz	8.5–10.68 GHz
Ku	12–18 GHz	13.4–14, 15.7–17.7 GHz
K	18–27 GHz	24.05–24.25 GHz
Ka	27–40 GHz	33.4–36 GHz
R	26.5–40 GHz	
Q	33–50 GHz	
V	40–75 GHz	
W	75–110 GHz	

[2–4]. Table 1.2 shows the nominal band designations and the official radar band designations in Region 2 as determined by international agreement through the International Telecommunications Union (ITU).

RF propagation modeling is still a maturing field as evidenced by the vast number of different models and the continual development of new models. Most propagation models considered in this text, while loosely based on physics, are empirical in nature. Wide variation in environments makes definitive models difficult, if not impossible, to achieve except in the simplest of circumstances, such as free-space propagation.

1.2 MODES OF PROPAGATION

Electromagnetic wave propagation is described by Maxwell's equations, which state that a changing magnetic field produces an electric field and a changing electric field produces a magnetic field. Thus electromagnetic waves are able to self-propagate. There is a well-developed theory on the subtleties of electromagnetic waves that is beyond the requirements of this book [5–7]. An introduction to the subject and some excellent references are provided in the second chapter. For most RF propagation modeling, it is sufficient to visualize the electromagnetic wave by a ray (the Poynting vector) in the direction of propagation. This technique is used throughout the book and is discussed further in Chapter 2.

1.2.1 Line-of-Sight Propagation and the Radio Horizon

In free space, electromagnetic waves are modeled as propagating outward from the source in all directions, resulting in a spherical wave front. Such a source is called an isotropic radiator and in the strictest sense, does not exist. As the distance from the source increases, the spherical wave (or phase) front converges to a planar wave front over any finite area of interest, which is how the propagation is modeled. The direction of propagation at any given point on the wave front is given by the vector cross product of the electric (**E**) field and the magnetic (**H**) field at that point. The *polarization* of a wave is defined as the orientation of the plane that contains the **E** field. This will be discussed further in the following chapters, but for now it is sufficient to understand that the polarization of the receiving antenna should ideally be the same as the polarization of the received wave and that the polarization of a transmitted wave is the same as that of the antenna from which it emanated.*

$$\mathbf{P} = \mathbf{E} \times \mathbf{H}$$

This cross product is called the Poynting vector. When the Poynting vector is divided by the characteristic impedance of free space, the resulting vector gives both the direction of propagation and the power density.

The power density on the surface of an imaginary sphere surrounding the RF source can be expressed as

$$S = \frac{P}{4\pi d^2} \quad \mathrm{W/m^2} \tag{1.1}$$

where d is the diameter of the imaginary sphere, P is the total power at the source, and S is the power density on the surface of the sphere in watts/m^2 or

---

* Neglecting any environmental effects.

equivalent. This equation shows that the power density of the electromagnetic wave is inversely proportional to d^2. If a fixed aperture is used to collect the electromagnetic energy at the receive point, then the received power will also be inversely proportional to d^2.

The velocity of propagation of an electromagnetic wave depends upon the medium. In free space, the velocity of propagation is approximately

$$c = 3 \times 10^8 \text{ m/s}$$

The velocity of propagation through air is very close to that of free space, and the same value is generally used. The *wavelength* of an electromagnetic wave is defined as the distance traversed by the wave over one cycle (period) and is generally denoted by the lowercase Greek letter lambda:

$$\lambda = \frac{c}{f} \tag{1.2}$$

The units of wavelength are meters or another measure of distance.

When considering line-of-sight (LOS) propagation, it may be necessary to consider the curvature of the earth (Figure 1.1). The curvature of the earth is a fundamental geometric limit on LOS propagation. In particular, if the distance between the transmitter and receiver is large compared to the height of the antennas, then an LOS may not exist. The simplest model is to treat the earth as a sphere with a radius equivalent to the equatorial radius of the earth.

From geometry

$$d^2 + r^2 = (r+h)^2$$

So

$$d^2 = (2r+h)h$$

and

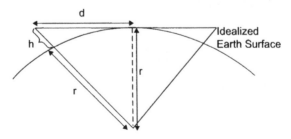

Figure 1.1 LOS propagation geometry over curved earth.

$$d \cong \sqrt{2rh} \qquad (1.3)$$

since $rh \gg h^2$.

The radius of the earth is approximately 3960 miles at the equator. The atmosphere typically bends horizontal RF waves downward due to the variation in atmospheric density with height. While this is discussed in detail later on, for now it is sufficient to note that an accepted means of correcting for this curvature is to use the "4/3 earth approximation," which consists of scaling the earth's radius by 4/3 [8]. Thus

$$r = 5280 \text{ miles}$$

and

$$d \cong \sqrt{2\frac{4}{3}3960\frac{h}{5280}}$$

or

$$\boxed{d \cong \sqrt{2h}} \qquad (1.4)$$

where d is the distance to the "radio horizon" in miles and h is in feet (5280 ft = 1 mi). This approximation provides a quick method of determining the distance to the radio horizon for each antenna, the sum of which is the maximum LOS propagation distance between the two antennas.

Example 1.1. Given a point-to-point link with one end mounted on a 100-ft tower and the other on a 50-ft tower, what is the maximum possible (LOS) link distance?

$$d_1 = \sqrt{2 \cdot 100} \cong 14.1 \text{ miles}$$
$$d_2 = \sqrt{2 \cdot 50} = 10 \text{ miles}$$

So the maximum link distance is approximately 24 miles. □

1.2.2 Non-LOS Propagation

There are several means of electromagnetic wave propagation beyond LOS propagation. The mechanisms of non-LOS propagation vary considerably, based on the operating frequency. At VHF and UHF frequencies, indirect propagation is often used. Examples of indirect propagation are cell phones, pagers, and some military communications. An LOS may or may not exist for these systems. In the absence of an LOS path, diffraction, refraction, and/or multipath reflections are the dominant propagation modes. *Diffraction* is the phenomenon of electromagnetic waves bending at the edge of a blockage, resulting in the shadow of the blockage being partially filled-in. *Refraction* is the bending of electromagnetic waves due to inhomogeniety in the medium. *Multipath* is the effect of reflections from multiple objects in the field of view, which can result in many different copies of the wave arriving at the receiver.

The over-the-horizon propagation effects are loosely categorized as *sky waves, tropospheric waves,* and *ground waves.* Sky waves are based on ionospheric reflection/refraction and are discussed presently. Tropospheric waves are those electromagnetic waves that propagate through and remain in the lower atmosphere. Ground waves include surface waves, which follow the earth's contour and space waves, which include direct, LOS propagation as well as ground-bounce propagation.

1.2.2.1 Indirect or Obstructed Propagation

While not a literal definition, indirect propagation aptly describes terrestrial propagation where the LOS is obstructed. In such cases, reflection from and diffraction around buildings and foliage may provide enough signal strength for meaningful communication to take place. The efficacy of indirect propagation depends upon the amount of margin in the communication link and the strength of the diffracted or reflected signals. The operating frequency has a significant impact on the viability of indirect propagation, with lower frequencies working the best. HF frequencies can penetrate buildings and heavy foliage quite easily. VHF and UHF can penetrate building and foliage also, but to a lesser extent. At the same time, VHF and UHF will have a greater tendency to diffract around or reflect/scatter off of objects in the path. Above UHF, indirect propagation becomes very inefficient and is seldom used. When the features of the obstruction are large compared to the wavelength, the obstruction will tend to reflect or diffract the wave rather than scatter it.

1.2.2.2 Tropospheric Propagation

The troposphere is the first (lowest) 10 km of the atmosphere, where weather effects exist. Tropospheric propagation consists of reflection (refraction) of RF from temperature and moisture layers in the atmosphere. Tropospheric propagation is less reliable than ionospheric propagation, but the phenomenon occurs often enough to be a concern in frequency planning. This effect is sometimes called *ducting*, although technically ducting consists of an elevated channel or duct in the atmosphere. Tropospheric propagation and ducting are discussed in detail in Chapter 6 when atmospheric effects are considered.

1.2.2.3 Ionospheric Propagation

The ionosphere is an ionized plasma around the earth that is essential to sky-wave propagation and provides the basis for nearly all HF communications beyond the horizon. It is also important in the study of satellite communications at higher frequencies since the signals must transverse the ionosphere, resulting in refraction, attenuation, depolarization, and dispersion due to frequency dependent group delay and scattering.

HF communication relying on ionospheric propagation was once the backbone of all long-distance communication. Over the last few decades, ionospheric propagation has become primarily the domain of shortwave broadcasters and radio amateurs. In general, ionospheric effects are considered to be more of a communication impediment rather than facilitator, since most commercial long-distance communication is handled by cable, fiber, or satellite. Ionospheric effects can impede satellite communication since the signals must pass through the ionosphere in each direction. Ionospheric propagation can sometimes create interference between terrestrial communications systems operating at HF and even VHF frequencies, when signals from one geographic area are scattered or refracted by the ionosphere into another area. This is sometimes referred to as *skip*.

The ionosphere consists of several layers of ionized plasma trapped in the earth's magnetic field (Figure 1.2) [9, 10]. It typically extends from 50 to 2000 km above the earth's surface and is roughly divided into bands (apparent reflective heights) as follows:

D	45–55 miles
E	65–75 miles
F1	90–120 miles
F2	200 miles (50–95 miles thick)

The properties of the ionosphere are a function of the free electron density, which in turn depends upon altitude, latitude, season, and primarily solar conditions.

Typically, the D and E bands disappear (or reduce) at night and F1 and F2 combine. For sky-wave communication over any given path at any given time there exists a *maximum usable frequency* (MUF) above which signals are no longer refracted, but pass through the F layer. There is also a *lowest usable*

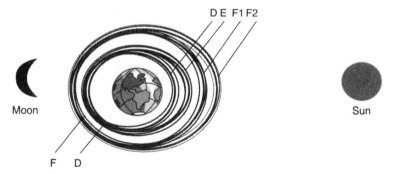

Figure 1.2   The ionospheric layers during daylight and nighttime hours.

frequency (LUF) for any given path, below which the D layer attenuates too much signal to permit meaningful communication.

The D layer absorbs and attenuates RF from 0.3 to 4 MHz. Below 300 kHz, it will bend or refract RF waves, whereas RF above 4 MHz will be passed unaffected. The D layer is present during daylight and dissipates rapidly after dark. The E layer will either reflect or refract most RF and also disappears after sunset. The F layer is responsible for most sky-wave propagation (reflection and refraction) after dark.

Faraday rotation is the random rotation of a wave's polarization vector as it passes through the ionosphere. The effect is most pronounced below about 10 GHz. Faraday rotation makes a certain amount of polarization loss on satellite links unavoidable. Most satellite communication systems use circular polarization since alignment of a linear polarization on a satellite is difficult and of limited value in the presence of Faraday rotation.

Group delay occurs when the velocity of propagation is not equal to c for a wave passing through the ionosphere. This can be a concern for ranging systems and systems that reply on wide bandwidths, since the group delay does vary with frequency. In fact the group delay is typically modeled as being proportional to $1/f^2$. This distortion of wideband signals is called *dispersion*. *Scintillation* is a form of very rapid fading, which occurs when the signal attenuation varies over time, resulting in signal strength variations at the receiver.

When a radio wave reaches the ionosphere, it can be refracted such that it radiates back toward the earth at a point well beyond the horizon. While the effect is due to refraction, it is often thought of as being a reflection, since that is the apparent effect. As shown in Figure 1.3, the point of apparent reflection is at a greater height than the area where the refraction occurs.

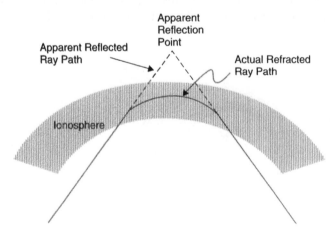

Figure 1.3   Geometry of ionospheric "reflection."

1.2.3 Propagation Effects as a Function of Frequency

As stated earlier, RF propagation effects vary considerably with the frequency of the wave. It is interesting to consider the relevant effects and typical applications for various frequency ranges.

The very low frequency (VLF) band covers 3–30 kHz. The low frequency dictates that large antennas are required to achieve a reasonable efficiency. A good rule of thumb is that the antenna must be on the order of one-tenth of a wavelength or more in size to provide efficient performance. The VLF band only permits narrow bandwidths to be used (the entire band is only 27 kHz wide). The primarily mode of propagation in the VLF range is ground-wave propagation. VLF has been successfully used with underground antennas for submarine communication.

The low-(LF) and medium-frequency (MF) bands, cover the range from 30 kHz to 3 MHz. Both bands use ground-wave propagation and some sky wave. While the wavelengths are smaller than the VLF band, these bands still require very large antennas. These frequencies permit slightly greater bandwidth than the VLF band. Uses include broadcast AM radio and the WWVB time reference signal that is broadcast at 60 kHz for automatic ("atomic") clocks.

The high-frequency (HF), band covers 3–30 MHz. These frequencies support some ground-wave propagation, but most HF communication is via sky wave. There are few remaining commercial uses due to unreliability, but HF sky waves were once the primary means of long-distance communication. One exception is international AM shortwave broadcasts, which still rely on ionospheric propagation to reach most of their listeners. The HF band includes citizens' band (CB) radio at 27 MHz. CB radio is an example of poor frequency reuse planning. While intended for short-range communication, CB signals are readily propagated via sky wave and can often be heard hundreds of miles away. The advantages of the HF band include inexpensive and widely available equipment and reasonably sized antennas, which was likely the original reason for the CB frequency selection. Several segments of the HF band are still used for amateur radio and for military ground and over-the-horizon communication.

The very high frequency (VHF) and ultra-high frequency (UHF) cover frequencies from 30 MHz to 3 GHz. In these ranges, there is very little ionospheric propagation, which makes them ideal for frequency reuse. There can be tropospheric effects, however, when conditions are right. For the most part, VHF and UHF waves travel by LOS and ground-bounce propagation. VHF and UHF systems can employ moderately sized antennas, making these frequencies a good choice for mobile communications. Applications of these frequencies include broadcast FM radio, aircraft radio, cellular/PCS telephones, the Family Radio Service (FRS), pagers, public service radio such as police and fire departments, and the Global Positioning System (GPS). These bands are the region where satellite communication begins since the signals can penetrate the ionosphere with minimal loss.

The super-high-frequency (SHF) frequencies include 3–30 GHz and use strictly LOS propagation. In this band, very small antennas can be employed, or, more typically, moderately sized directional antennas with high gain. Applications of the SHF band include satellite communications, direct broadcast satellite television, and point-to-point links. Precipitation and gaseous absorption can be an issue in these frequency ranges, particularly near the higher end of the range and at longer distances.

The extra-high-frequency (EHF) band covers 30–300 GHz and is often called *millimeter wave*. In this region, much greater bandwidths are available. Propagation is strictly LOS, and precipitation and gaseous absorption are a significant issue.

Most of the modeling covered in this book is for the VHF, UHF, SHF, and lower end of the EHF band. VHF and UHF work well for mobile communications due to the reasonable antenna sizes, minimal sensitivity to weather, and moderate building penetration. These bands also have limited over-the-horizon propagation, which is desirable for frequency reuse. Typical applications employ vertical antennas (and vertical polarization) and involve communication through a centrally located, elevated repeater.

The SHF and EHF bands are used primarily for satellite communication and point-to-point communications. While they have greater susceptibility to environmental effects, the small wavelengths make very high gain antennas practical.

Most communication systems require two-way communications. This can be accomplished using half-duplex communication where each party must wait for a clear channel prior to transmitting. This is sometimes called carrier-sensed multiple access (CSMA) when done automatically for data communications, or push-to-talk (PTT) in reference to walkie-talkie operation. Full duplex operation can be performed when only two users are being serviced by two independent communication channels, such as when using frequency duplexing.* Here each user listens on the other user's transmit frequency. This approach requires twice as much bandwidth but permits a more natural form of voice communication. Other techniques can be used to permit many users to share the same frequency allocation, such as time division multiple access (TDMA) and code division multiple access (CDMA).

1.3   WHY MODEL PROPAGATION?

The goal of propagation modeling is often to determine the probability of satisfactory performance of a communication system or other system that is dependent upon electromagnetic wave propagation. It is a major factor in communication network planning. If the modeling is too conservative, excessive costs may be incurred, whereas too liberal of modeling can result in

---

* Sometimes called two-frequency simplex.

unsatisfactory performance. Thus the fidelity of the modeling must fit the intended application.

For communication planning, the modeling of the propagation channel is for the purpose of predicting the received signal strength at the end of the link. In addition to signal strength, there are other channel impairments that can degrade link performance. These impairments include *delay spread* (smearing in time) due to multipath and rapid signal fading within a symbol (distortion of the signal spectrum). These effects must be considered by the equipment designer, but are not generally considered as part of communication link planning. Instead, it is assumed that the hardware has been adequately designed for the channel. In some cases this may not hold true and a communication link with sufficient receive signal strength may not perform well. This is the exception rather than the norm however.

1.4 MODEL SELECTION AND APPLICATION

The selection of the model to be used for a particular application often turns out to be as much art (or religion) as it is science. Corporate culture may dictate which models will be used for a given application. Generally, it is a good idea to employ two or more independent models if they are available and use the results as bounds on the expected performance. The process of propagation modeling is necessarily a statistical one, and the results of a propagation analysis should be used accordingly.

There may be a temptation to "shop" different models until one is found that provides the desired answer. Needless to say, this can lead to disappointing performance at some point in the future. Even so, it may be valuable for certain circumstances such as highly competitive marketing or proposal development. It is important that the designer not be lulled into placing too much confidence in the results of a single model, however, unless experience shows it to be a reliable predictor of the propagation channel that is being considered.

1.4.1 Model Sources

Many situations of interest have relatively mature models based upon large amounts of empirical data collected specifically for the purpose of characterizing propagation for that application. There are also a variety of proprietary models based on data collected for very specific applications. For more widely accepted models, organizations like the International Telecommunications Union (ITU) provide recommendations for modeling various types of propagation impairments. While these models may not always be the best suited for a particular application, their wide acceptance makes them valuable as a benchmark.

There exist a number of commercially available propagation modeling software packages. Most of these packages employ standard modeling techniques.

In addition, some may include proprietary models. When using such packages, it is important that the user have an understanding of what the underlying models are and the limitations of those models.

1.5 SUMMARY

In free space, the propagation loss between a transmitter and receiver is readily predicted. In most applications however, propagation is impaired by proximity to the earth, objects blocking the LOS and/or atmospheric effects. Because of these impairments, the fundamental characteristics of RF propagation generally vary with the frequency of the electromagnetic wave being propagated. The frequency spectrum is grouped into bands, which are designated by abbreviations such as HF, VHF, and so on. Letter designations of the bands are also used, although the definitions can vary.

Propagation of electromagnetic waves may occur by ground wave, tropospheric wave, or sky wave. Most contemporary communication systems use either direct LOS or indirect propagation, where the signals are strong enough to enable communication by reflection, diffraction, or scattering. Ionospheric and tropospheric propagation are rarely used, and the effects tend to be treated as nuisances rather than a desired means of propagation.

For LOS propagation, the approximate distance to the apparent horizon can be determined using the antenna height and the 4/3's earth model. Propagation effects tend to vary with frequency, with operation in different frequency bands sometimes requiring the designer to address different phenomena. Modeling propagation effects permits the designer to tailor the communication system design to the intended environment.

REFERENCES

1. J. D. Parsons, *The Mobile Radio Propagation Channel*, 2nd ed., John Wiley & Sons, West Sussex, UK, 2000, Table 1.1.
2. M. I. Skolnik, *Introduction to Radar Systems*, 3rd ed., McGraw-Hill, New York, 2001, Table 1.1, p. 12.
3. L. W. Couch II, *Digital and Analog Communication Systems*, 6th ed., Prentice-Hall, Upper Saddle River, NJ, 2001, Table 1.2.
4. ITU Recommendations, Nomenclature of the frequency and wavelength bands used in telecommunications, ITU-R V.431-6.
5. M. A. Plonus, *Applied Electromagnetics*, McGraw-Hill, New York, 1978.
6. S. Ramo, J. R. Whinnery, and T. Van Duzer, *Fields and Waves in Communication Electronics*, 2nd ed., John Wiley & Sons, New York, 1984.
7. S. V. Marshall and G. G. Skitek, *Electromagnetic Concepts & Applications*, 2nd ed., Prentice-Hall, Englewood Cliffs, NJ, 1987.

8. M. I. Skolnik, *Introduction to Radar Systems*, 3rd ed., McGraw-Hill, New York, 2001, p. 496.

9. L. W. Couch II, *Digital and Analog Communication Systems*, 6th ed., Prentice-Hall, Upper Saddle River, NJ, 2001, pp. 12–16.

10. *The ARRL Handbook for Radio Amateurs*, ARRL, Newington, CT, 1994, pp. 22-1–22-15.

EXERCISES

1. Determine the following;
   (a) What is the wavelength of an 800-MHz signal?
   (b) What is the wavelength of a 1.9-GHz signal?
   (c) What is the wavelength of a 38-GHz signal?
   (d) What is the frequency of a 10-m amateur radio signal?
   (e) What effect would the ionosphere have on signals at each of the preceding frequencies?

2. What is the power density of a 1-kW signal radiated from an isotropic radiator, at a distance of 10 km?

3. Determine the following;
   (a) What is the distance to the horizon as viewed from a height of 1 m and from 10 m?
   (b) What is the distance to the radio horizon for the same heights
   (c) What is the maximum LOS communication distance (i.e., neglecting any ground-wave or ionospheric propagation) between two systems, one with the antenna mounted 1 m above the ground and the other with the antenna mounted 10 m above the ground?

CHAPTER 2

Electromagnetics and RF Propagation

2.1 INTRODUCTION

This chapter provides a brief review of electromagnetic theory. While not exhaustive, it provides sufficient background and review for understanding the material in later chapters. The discussions include the concepts of electric and magnetic fields, the wave equation, and electromagnetic wave polarization. The physics behind reflection and refraction of electromagnetic waves is discussed and used to make some generalizations about RF ground reflections.

There are two orthogonal time-varying fields that comprise an electromagnetic plane wave: electric and magnetic. Each has its own distinct properties; when related by the wave equation, they form the mathematical basis for electromagnetic wave propagation. Prior to examining the wave equation and its implications for electromagnetic wave propagation, it is worthwhile to take a brief look at static electric and magnetic fields.

2.2 THE ELECTRIC FIELD

The unit of electric charge is the coulomb. The electric field is generated by an electric charge and is defined as the vector force exerted on a unit charge and is usually denoted by $\mathbf{E}$. Thus the units of electric field are newtons per coulomb, which is equivalent to volts per meter.* The electric field is dependent upon the amount of flux, or the *flux density* and the permittivity, ε, of the material:

$$\mathbf{E} = \varepsilon \mathbf{D}$$

---

* In SI units, a volt is equal to a $kg \cdot m^2/(A \cdot s^2)$, a coulomb is equal to an $A \cdot s$, and a newton is a $kg \cdot m/s^2$.

Introduction to RF Propagation, by John S. Seybold
Copyright © 2005 by John Wiley & Sons, Inc.

The flux density vector is defined as a vector that has the same direction as the **E** field and whose strength is proportional to the charge that generates the **E** field. The units of the flux density vector are coulombs per square meter. The **E** field and the flux density vectors originate at positive charges and terminate on negative charges or at infinity as shown in Figure 2.1.

Gauss's Law states that the total charge within an enclosed surface is equal to the integral of the flux density over that surface. This can be useful for mathematically solving field problems when there is symmetry involved. The classic field problems are the electric field from a point charge, an infinite line charge and an infinite plane or surface charge. These are shown in Table 2.1.

2.2.1 Permittivity

Since the electric or **E** field depends not only on the flux density, but also on the permittivity of the material or environment through which the wave is propagating, it is valuable to have some understanding of permittivity. Permittivity is a property that is assigned to a dielectric (conductors do not support static electric fields). The permittivity is a metric of the number of bound charges in a material [1] and has units of farads per meter. For mathematical tractability, uniform (homogeneous) and time-invariant permittivity is assumed throughout this chapter. Permittivity is expressed as a multiple of the permittivity of free space, ε_0. This term is called the *relative permittivity*, ε_r, or the *dielectric constant* of the material.

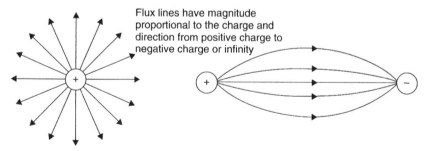

Figure 2.1   Diagram of **E** field and flux density vectors about an electric charge.

TABLE 2.1   Electric Field Intensity for Classic E-Field Geometries

E Field Source	Symmetry	Field Intensity
Point charge	Spherical	$\mathbf{E} = \dfrac{Q}{4\pi r^2 \varepsilon}\hat{\mathbf{r}}$
Infinite line charge	Cylindrical	$\mathbf{E} = \dfrac{Q}{2\pi r \varepsilon}\hat{\mathbf{r}}$
Infinite surface charge	Planar	$\mathbf{E} = Q\hat{\mathbf{z}}$

$$\varepsilon = \varepsilon_r \varepsilon_0$$

with

$$\varepsilon_0 = 8.854 \times 10^{-12} \text{ F/m}$$

Table 2.2 gives some representative values of dielectric constant for some common materials.

TABLE 2.2   Relative Permittivity of Some Common Materials

Material	Relative Permittivity
Vacuum	1
Air	1.0006
Polystyrene	2.7
Rubber	3
Bakelite	5
Quartz	5
Lead glass	6
Mica	6
Distilled water	81

Source: Plonus [2], courtesy of McGraw-Hill.

The boundary between two dissimilar dielectrics will bend or refract an electric field vector.* This is due to the fact that the component of the *flux density vector* that is normal to the boundary is constant across the boundary, while the parallel component of the *electric field* is constant at the boundary. This is shown in Figure 2.2 where the N and T subscripts denote the normal and tangential components of the electric field vector and flux density vectors relative to the dielectric boundary, respectively. The subtleties of why this occurs is treated in the references [3, 4]. Thus

$$\phi_1 = \tan^{-1}\left(-\vec{E}_{N1}/\vec{E}_{T1}\right)$$

and

$$\phi_2 = \tan^{-1}\left(\vec{E}_{N2}/\vec{E}_{T2}\right)$$

Since

$$D_{N1} = D_{N2}$$

it is apparent that

---

* This is a bending of the **E** field vector, which is different than the bending of the propagation vector that occurs when a wave is incident on a dielectric boundary.

$$\vec{E}_{N2} = \frac{\varepsilon_2}{\varepsilon_1} \vec{E}_{N1}$$

Using the fact that

$$E_{T1} = E_{T2}$$

and substituting into the expression for ϕ_2, yields

$$\phi_2 = \tan^{-1}\left(\frac{\varepsilon_2}{\varepsilon_1} \frac{\vec{E}_{N1}}{\vec{E}_{T1}}\right) \tag{2.1}$$

Thus the angle or direction of the **E** field in the second dielectric can be written in terms of the components of the **E** field in the first dielectric and the ratio of the dielectric constants. Taking the derivation one step further, the ratio of E_{N1} to E_{T1} is the tangent of ϕ_1, so

$$\boxed{\phi_2 = \tan^{-1}\left(\frac{\varepsilon_2}{\varepsilon_1} \tan(\phi_1)\right)} \tag{2.2}$$

This expression shows that, as one might expect, electric fields that are normal or parallel to a dielectric boundary ($\phi_1 = 0$ or $90°$) are not refracted and the amount of refraction depends upon the dielectric constants of both materials and the grazing angle.

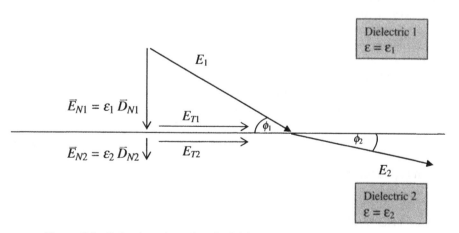

Figure 2.2 Behavior of an electric field vector at a dielectric boundary.

2.2.2 Conductivity

Materials that have free electrons available are called *conductors*. Conductors are characterized by their conductivity, σ, or by the reciprocal of the conduc-

TABLE 2.3   Some Representative Values for Conductivity for Various Materials Ranging from Conductors to Insulators (From [5], courtesy of McGraw-Hill)

Conductors		Insulators	
Material	Conductivity (S/m)	Material	Conductivity (S/m)
Silver	6.1×10^7	Wet earth	$\sim 10^{-3}$
Copper	5.7×10^7	Silicon	3.9×10^{-4}
Gold	4.1×10^7	Distilled water	$\sim 10^{-4}$
Aluminum	3.5×10^7	Dry earth	$\sim 10^{-5}$
Tungsten	1.8×10^7	Rock	$\sim 10^{-6}$
Brass	1.1×10^7	Bakelite	$\sim 10^{-9}$
Iron	$\sim 10^7$	Glass	$\sim 10^{-12}$
Nichrome	10^6	Rubber	$\sim 10^{-15}$
Mercury	10^6	Mica	$\sim 10^{-15}$
Graphite	10^5	Wax	$\sim 10^{-17}$
Carbon	3×10^7	Quartz	$\sim 10^{-17}$
Germanium	2.3		
Seawater	4		

Source: Plonus [5], courtesy of McGraw-Hill.

tivity, which is the resistivity, ρ. The units of conductivity are siemens per meter, and the units of resistivity are ohm-meters. Materials with very low conductivity are called *insulators*. A perfect dielectric will have zero conductivity, while most real-world materials will have both a dielectric constant and a non-zero conductivity. As the conductivity of the dielectric material increased, the dielectric becomes more lossy. When considering the effect of nonideal materials on electromagnetic waves, the permittivity can be expressed as a complex number that is a function of the dielectric constant, the conductivity, and the frequency of the wave. This is discussed further in Section 2.4.2.

A static electric field cannot exist in a conductor, because the free electrons will move in response to the electric field until it is balanced. Thus, when an electric field is incident on a conductor, enough free electrons will move to the surface of the conductor to balance the incident electric field, resulting in a surface charge on the conductor. This phenomenon is central to the operation of a capacitor [6].

2.3   THE MAGNETIC FIELD

Static magnetic fields can be generated by steady (or linearly increasing) current flow or by magnetic materials. The magnetic field has both strength and direction, so it is denoted by a vector, **B**. In a manner similar to the electric field, the magnetic field can be divided into magnetic flux density (**H**) and magnetic field strength (**B**). The unit of magnetic field strength is amperes per meter (A/m), and the unit of magnetic flux density is webers per square meter (Wb/m²) or teslas. For nonmagnetic materials, the magnetic field and the magnetic flux density are linearly related by, μ, the permeability of the material.

$$\mathbf{B} = \mu\mathbf{H}$$

The units of material permeability are henries per meter where the henry is the unit of inductance. One henry equals one weber per ampere. The permeability is expressed as a relative permeability, μ_r, times the permeability of free space μ_0, so

$$\mu = \mu_r\mu_0$$

where

$$\mu_0 = 4\pi \times 10^{-7} \; H/m$$

The magnetic flux is proportional to current flow, while the magnetic field depends on both the current flow and the permeability of the material. The Biot–Savart Law (alternately known as Ampere's Law) quantifies the relationship between electrical current and magnetic flux. In order to have a steady current flow, a closed circuit is required. Figure 2.3 shows a representative geometry. The integral form of the Biot–Savart Law is

$$\overline{H} = \oint \frac{\overline{Idl} \times \hat{a}_R}{4\pi R^2} \quad A/m$$

which says that the magnetic field is equal to the line integral of the current crossed with the vector from the current loop to the point of interest and scaled by 4π times the separation squared.

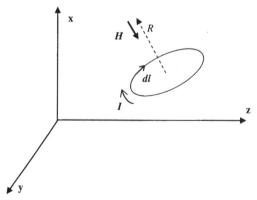

Figure 2.3   Geometry of a current loop and the resulting magnetic field.

When a magnetic field is incident on a boundary between magnetic materials, the relevant equations are

$$\mathbf{B}_{N1} = \mathbf{B}_{N2}$$

and

$$\mathbf{H}_{T1} - \mathbf{H}_{T2} = \mathbf{J}_S$$

where $\mathbf{J}_S$ is the surface current density vector at the boundary.

For the purposes of this book the conductive and dielectric properties are of primary interest, and it will usually be assumed that the magnetic permeability of the materials being considered is unity ($\mu_r = 1$) unless otherwise specified.

2.4 ELECTROMAGNETIC WAVES

Maxwell's equations form the basis of electromagnetic wave propagation. The essence of Maxwell's equations are that a time-varying electric field produces a magnetic field and a time-varying magnetic field produces an electric field. A time-varying magnetic field can be generated by an accelerated charge.

In this book the focus is on plane waves, since plane waves represent electromagnetic radiation at a distance from the source and when there are no interfering objects in the vicinity. In a strict sense, all real waves are spherical, but at a sufficient distance from the source, the spherical wave can be very well approximated by a plane wave with linear field components, over a limited extent. When using plane waves, the electric field, magnetic field, and direction of propagation are all mutually orthogonal. By using the propagation direction vector to represent the plane wave, the visualization and analysis of plane-wave propagation is greatly simplified. This is called *ray theory* and is used extensively in later chapters. Ray theory is very useful in far-field (plane-wave analysis), but is not universally applicable in near-field situations. The relationship between the time-varying electric and magnetic fields is expressed mathematically for uniform plane waves as

$$\nabla \times \overline{E} = -\mu \frac{\partial \overline{H}}{\partial t}$$

$$\nabla \times \overline{H} = \varepsilon \frac{\partial \overline{E}}{\partial t}$$

The differential form of Maxwell's equations are used to derive the wave equation [7], which is expressed in one dimension as

$$\frac{\partial^2 E_x}{\partial t^2} = \mu\varepsilon \frac{\partial^2 E_x}{\partial z^2}$$

This partial differential equation is the fundamental relationship that governs the propagation of electromagnetic waves. The velocity of propagation for the electromagnetic wave is determined from the wave equation and is a function of the permittivity and permeability of the medium.

$$v = \frac{1}{\sqrt{\mu\varepsilon}} \tag{2.3}$$

When expressed in terms of the relative permittivity and permeability, the equation for the velocity of propagation becomes

$$v = \frac{1}{\sqrt{\mu_r\varepsilon_r}\sqrt{\mu_0\varepsilon_0}}$$

Using the values of μ_0 and ε_0 in this expression yields

$$v = \frac{1}{\sqrt{\mu_r\varepsilon_r}}c \tag{2.4}$$

where

$$c = 2.998 \times 10^8 \text{ m/s}$$

Thus the velocity of propagation is equal to the velocity of light in free space divided by the square root of the product of the relative permittivity and permeability.

An electromagnetic plane wave traveling in the positive z direction can be described by the following equations:

$$P(z) = E_x(z)H_y(z)$$

where

$$E_x(z) = E_1 e^{-jkz} + E_2 e^{jkz}$$
$$\eta H_y(z) = E_1 e^{-jkz} + E_2 e^{jkz}$$

and k is the wave number,

$$k = \omega\sqrt{\mu\varepsilon} = \frac{2\pi}{\lambda} \text{ m}^{-1} \tag{2.5}$$

Note that a factor of $e^{-j\omega t}$ has been suppressed in the expressions for E_x and H_y.

2.4.1 Electromagnetic Waves in a Perfect Dielectric

The form of electromagnetic waves in a uniform lossless dielectric is identical to the form in free space with the exception of the value of permittivity. Thus by simply multiplying the free-space electric field component by the dielectric constant of the material, the expression for the plane wave in the dielectric is obtained. In each case, the relative permeability is assumed to be equal to one.

2.4.2 Electromagnetic Waves in a Lossy Dielectric or Conductor

In a practical sense, most materials lie on a continuum of properties. The characterization of a material as a conductor or dielectric is based on the dominant property of the material. For lossy dielectrics, the permittivity or dielectric constant is given by

$$\varepsilon = \varepsilon'\left(1 - \frac{j\sigma}{\omega\varepsilon'}\right) \tag{2.6}$$

where σ is the conductivity of the dielectric. Thus the dielectric constant is a complex value with the imaginary part representing the loss characteristics of the material. The loss tangent is defined as $\sigma/(\omega\varepsilon')$ and represents the ratio of conductive current to displacement current in the material. A material can be considered low loss if the loss tangent is less than 0.1, and it is considered high loss if the loss tangent is greater than 10.

2.4.3 Electromagnetic Waves in a Conductor

When a static electric field is incident on a conductor, the free charges in the conductor simply move to cancel the electric field. In the case of an electromagnetic wave, however, the incident field is changing, so the currents in the conductor are also changing, which produces an electromagnetic wave. Thus an electromagnetic wave can exist within a conductor, although it will be concentrated near the surface. By incorporating the complex permittivity into the wave equation, it is clear that the complex part of the permittivity produces a real value in the exponent, which causes the wave strength to decay with distance as it penetrates the conducting material. To see this, consider the complex wave number,

$$k = \omega\sqrt{\mu\varepsilon} = \omega\sqrt{\mu\varepsilon'\left(1 - \frac{j\sigma}{\omega\varepsilon}\right)} \qquad \mathrm{m}^{-1} \tag{2.7a}$$

or

$$k = \omega \sqrt{\mu \left(\varepsilon' - \frac{j\sigma}{\omega'} \right)} \quad \mathrm{m}^{-1} \qquad (2.7b)$$

The wave number can then be separated into real and imaginary parts [8]:

$$jk = \alpha + j\beta = j\omega \sqrt{\mu \left(\varepsilon' - \frac{j\sigma}{\omega'} \right)} \quad \mathrm{m}^{-1} \qquad (2.8)$$

The expression for α, the attenuation constant can be shown to be

$$\alpha = \omega \sqrt{\frac{\mu \varepsilon'}{2} \left(\sqrt{1 + \left(\frac{\sigma}{\omega \varepsilon'} \right)^2} - 1 \right)} \qquad (2.9a)$$

and the expression for β, the phase constant is

$$\beta = \omega \sqrt{\frac{\mu \varepsilon'}{2} \left(\sqrt{1 + \left(\frac{\sigma}{\omega \varepsilon'} \right)^2} + 1 \right)} \qquad (2.9b)$$

Thus, when an electromagnetic wave is incident on a conductive material, it penetrates the material according to the following equations [9]:

$$E_x(z) = E_1 e^{-(a+j\beta)z} + E_2 e^{(a+j\beta)z}$$

$$\eta H_y(z) = E_1 e^{-(a+j\beta)z} + E_2 e^{(a+j\beta)z}$$

Thus the real part of the exponent is called the *attenuation constant*, and it causes the amplitude of the wave to attenuate with distance (z). For relatively good conductors, the conductivity will be large and α can be approximated as

$$\alpha \cong \sqrt{\frac{\omega \mu \sigma}{2}} \qquad (2.10)$$

Since the wave is attenuated as it penetrates the conductor, it is useful to be able to provide a quick characterization of how deep the electromagnetic wave will penetrate into the conductive material. The attenuation constant is a function of the frequency of the wave, the conductivity of the material (as well as the permeability), and the distance traversed in the conducting material. The distance where the amplitude of the incident wave is attenuated by a factor of e^{-1} is called the *skin depth*. The skin depth is found by setting $\alpha \delta = 1$ and solving for δ. The result is

$$\delta = \frac{1}{\sqrt{\pi f \mu \sigma}} \quad \mathrm{m} \qquad (2.11)$$

The skin depth is frequently used to provide an assessment of the depth of the electromagnetic wave into a conductor. This can be useful, for example, when determining the required thickness of electromagnetic shielding.

2.5 WAVE POLARIZATION

The polarization of an electromagnetic wave is defined as the orientation of plane in which the electric field resides. Antenna polarization is defined as the polarization of the wave that it *transmits*. The simplest polarization to envision is linear, which is usually either vertical or horizontal polarization, but it can be defined for other orientations (Figure 2.4).

The vector cross product between the electrical and magnetic fields gives a vector in the direction of propagation. This is called the *Poynting vector* and can be defined as either

$$\mathbf{S} = \mathbf{E} \times \mathbf{H}$$

or

$$\mathbf{S} = \frac{1}{2}\mathbf{E} \times \mathbf{H}/Z_0 \qquad \mathrm{W/m^2}$$

where Z_0 is the *characteristic impedance* (or *intrinsic impedance*) of the medium, which is given by

$$Z_0 = \sqrt{\frac{\mu}{\varepsilon}} \qquad \mathrm{ohms}$$

which can also be expressed as

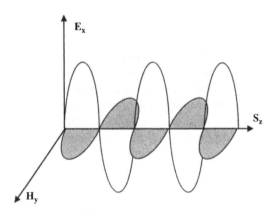

Figure 2.4 Conceptual diagram of linear polarization.

$$Z_0 = 377\sqrt{\mu_r / \varepsilon_r} \qquad \text{ohms} \qquad (2.12)$$

Thus the second formulation of the Poynting vector gives the power density in watts per square meter. The concept of power density is discussed further in later chapters.

Another type of polarization is circular, which is a special case of elliptical polarization. This is covered in more detail in the following chapter on antennas, but for now it is sufficient to note that elliptical polarization consists of the sum of two orthogonal, linearly polarized waves (usually vertical and horizontal) that are 90 degrees out of phase. The sign of the 90-degree phase difference sets the direction of the polarization, right hand or left hand. The *axial ratio* is defined as the ratio of the major axis of the ellipse to the minor axis. For circular polarization, the axial ratio is unity (0 dB). Chapter 3 provides more information on axial ratio.

In previous sections the effect of the material boundaries on magnetic and electric fields was discussed. The angle of incidence* of the electric and magnetic field along with the ratio of relative permittivity and permeability determine what the net effect of the material boundary will be. For this reason, the polarization of the wave as well as the angle of incidence (or the grazing angle) and material properties must be known in order to predict the effect of the boundary on the magnitude and direction of the resulting wave. This has interesting implications for elliptical polarization, where the sense of the linear polarizations are rotating. In this case, it is possible that some material boundaries may actually disperse or (linearly) polarize the incident wave rather than simply refracting and reflecting it.

2.6 PROPAGATION OF ELECTROMAGNETIC WAVES AT MATERIAL BOUNDARIES

For the purposes of RF propagation, the effect of the interaction of a plane wave with other surfaces such as knife-edge diffractors (see Chapter 8) or the ground are of considerable interest. In this chapter the effect of a plane wave incident on a flat (smooth) surface of either a perfect dielectric or a perfect conductor is characterized from a theoretical standpoint. The lossy-dielectric case is not treated in detail because the development is somewhat tedious and the material properties are not usually that well known in the environments of interest. In addition, the real-world problem of irregular surfaces and nonhomogeneous materials is not readily treated mathematically and for the

---

* The angle of incidence is defined as the angle relative to the surface normal. For work in this text, the grazing angle (the angle relative to the surface) is usually of greater interest and for that reason is used throughout. The conversion of the equations found in the electromagnetic literature is straightforward.

purpose of RF propagation analysis is generally treated using empirical data about the materials and/or surfaces in question.

When an electromagnetic wave is incident on a surface, some of the energy will be reflected and/or scattered and some will be absorbed/refracted. Scattering is reflection from sharp edges or irregular surfaces and is not treated in this chapter, since extensive detail about the surface is required to do so and a statistical characterization is usually preferred. When a wave is incident on a material surface, some of the energy will (generally) enter the material and the rest will be reflected. Of the energy that enters the material, some of it may be absorbed due to the conductivity of the material and the rest propagates into the material (is refracted). With the exception of circumstances such as penetrating walls or windows, the reflected component is of primary interest to the analyst.

2.6.1 Dielectric to Dielectric Boundary

The analysis of the reflection and refraction of an electromagnetic wave at a material boundary is based on applying appropriate boundary conditions to the plane-wave equation. One interesting and useful approach to this analysis is to treat the problem as a transmission line system [9]. By so doing, the details of the analysis can be concisely incorporated into a relatively few number of equations. This approach is used in the following development. An important subtlety of refraction of electromagnetic waves is that the refraction is taken to be the change in the direction of wave propagation, whereas for static electric and magnetic fields the refraction is taken as the change in the direction of the flux lines.

Prior to looking at the impact of electromagnetic waves on material boundaries, it is important to define the polarization angle relative to the plane of incidence, since the boundary effects are acting on the electric field portion of the wave since the materials considered herein are nonmagnetic. Figure 2.5 shows an electromagnetic wave incident on a boundary between two materi-

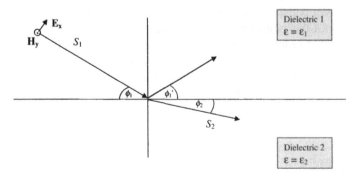

Figure 2.5 Diagram of a TM electromagnetic wave incident on a material boundary.

als. The electric field component of S_1 is designated by the small arrow, and the magnetic field is shown coming out of the page by the circle with the dot inside. This is formally called a *transverse magnetic* or TM wave, since the magnetic field is normal to the plane of incidence and the electric field is contained in the plane of incidence. The plane of incidence is defined as the plane formed by the propagation direction vector, S_1, and the material boundary. A more familiar example of a TM wave would be a vertically polarized wave incident on a horizontal surface. A transverse electric or TE wave would have the electric field coming out of the paper and the magnetic field pointing downward from S_1. A TE wave would correspond to a horizontally polarized wave incident on a horizontal surface or material boundary.

Figure 2.5 shows how, in general, an electromagnetic wave that is incident on a material boundary will have both a reflected and a refracted or transmitted component. This separation of the incident wave is treated by using the concept of a reflection and a transmission coefficient. It can be shown [10] that for a smooth, flat surface, the angle of reflection is equal to the grazing angle.

$$\phi_1' = \phi_1$$

Snell's law of refraction from optics provides another convenient relationship,*

$$\frac{\cos(\theta_2)}{\cos(\theta_1)} = \frac{v_2}{v_1} = \frac{n_1}{n_2} \tag{2.13}$$

where v_2 and v_1 are the propagation velocities in materials 2 and 1, respectively, and n_1 and n_2 are the refraction indexes ($n = c/v$) of materials 1 and 2, respectively. This may be expressed in terms of permittivity and permeability as

$$\frac{\cos(\theta_2)}{\cos(\theta_1)} = \frac{\sqrt{\mu_1/\varepsilon_1}}{\sqrt{\mu_{21}/\varepsilon_2}} = \sqrt{\frac{\varepsilon_2 \, \mu_1}{\varepsilon_1 \, \mu_2}} = \sqrt{\frac{\varepsilon_{r2} \, \mu_{r1}}{\varepsilon_{r1} \, \mu_{r2}}}$$

For most cases of interest, the relative permeability is unity and

$$\frac{\cos(\theta_2)}{\cos(\theta_1)} = \sqrt{\frac{\varepsilon_{r2}}{\varepsilon_{r1}}} \tag{2.14}$$

---

* It should be noted that electromagnetic texts define the angle of incidence to the normal vector at the surface rather than relative to the surface. The use of grazing angle is better suited toward the applications of RF analysis, but results in some of the equations being different from those in an electromagnetics book.

Note that (2.14) is different from (2.2). The difference is that (2.2) applies to the bending of the **E** field lines, whereas (2.14) applies to the direction of propagation, which is orthogonal to the **E** field. Put another way, ϕ is the angle of incidence of the electric field lines, and θ is the angle of incidence of the direction of travel of a wave. This is a subtle but important distinction.

The interaction of an electromagnetic wave with a boundary between two materials can be treated as a transmission line problem. The goal is to determine the reflection coefficient, ρ, and the transmission coefficient, τ. The expressions are slightly different depending upon the nature of the materials and the sense of the wave polarization. Referring to Figure 2.5, for flat surfaces, it will always be the case that the angle of departure for the reflected part of the wave will be equal to the grazing angle,

$$\phi_1 = \phi_1'$$

In addition, the following relationships hold:

$$\rho = \frac{Z_L - Z_{z1}}{Z_L + Z_{z1}} \tag{2.15}$$

$$\tau = \frac{2Z_L}{Z_L + Z_{z1}} \tag{2.16}$$

where Z_L is the effective wave impedance of the second material, which is a function of the grazing angle and the propagation velocities in each material.

$$\text{TM:} \quad Z_L = Z_2 \sin(\theta_2) = Z_2 \sqrt{1 - \left(\frac{v_2}{v_1}\right)^2 \cos^2(\theta_1)} \tag{2.17}$$

$$\text{TE:} \quad Z_L = Z_2 \csc(\theta_2) = Z_2 \left[1 - \left(\frac{v_2}{v_1}\right)^2 \cos^2(\theta_1)\right]^{-1/2} \tag{2.18}$$

Since the definition of the propagation velocity is given by (2.3), the following substitution can be made in (2.17) and (2.18):

$$\left(\frac{v_2}{v_1}\right)^2 = \frac{\varepsilon_1}{\varepsilon_2}$$

Either of the expressions for effective wave impedance may be used to determine the angle of the transmitted portion of the incident wave.

$$\theta_2 = \cos^{-1}\left[\sqrt{1 - \frac{\varepsilon_1}{\varepsilon_2}\sin^2(\theta_1)}\right] \tag{2.19}$$

A plot of the transmission angle versus grazing angle for several values of $\varepsilon_2/\varepsilon_1$ is given in Figure 2.6. The expression for the wave impedances in the first material are, of course, dependent upon the polarization and given by

$$\text{TM:} \qquad Z_{z1} = Z_1\sin(\theta_1) \qquad\qquad (2.20)$$

$$\text{TE:} \qquad Z_{z1} = Z_1\csc(\theta_1) \qquad\qquad (2.21)$$

The reflection and transmission coefficients for an electromagnetic wave incident on a perfect dielectric can now be calculated. Figure 2.7 shows plots of the reflection and transmission coefficients for the wave incident on a perfect dielectric. The values of, ε_r may be interpreted as the ratio $\varepsilon_2/\varepsilon_1$ or as simply ε_{r2} if the first material is free space. Several interesting observations can be made from these plots: The transmission coefficient minus the reflection coefficient is always equal to one, it is possible for the transmission coefficient to exceed one since the electric field will be stronger in the second material (if it has the higher dielectric constant), and when the dielectric constants of the two materials are equal, there is no reflection and total transmission of the wave as expected.

It is also noteworthy that there is a point for TM waves (i.e., vertical polarization) where the reflection coefficient crosses zero. The angle where this occurs is called the *critical angle*, and at that point there is total transmission of the wave and no reflection. The critical angle occurs when $Z_L = Z_{z1}$, which gives

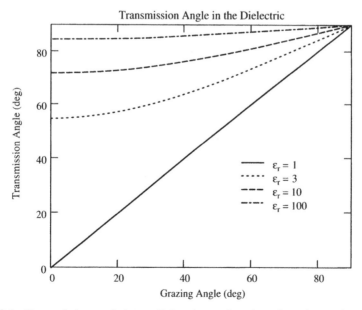

Figure 2.6   Transmission angle into a dielectric as a function of grazing angle and $\varepsilon_2/\varepsilon_1$.

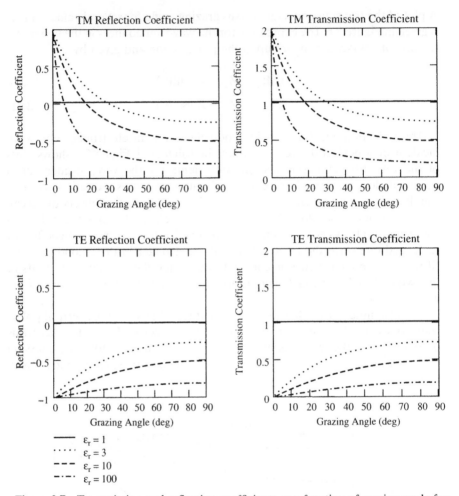

Figure 2.7 Transmission and reflection coefficients as a function of grazing angle for different values of $\varepsilon_2/\varepsilon_1$ at a boundary between perfect dielectrics.

$$Z_1 \sin(\theta_c) = Z_2 \sqrt{1 - \left(\frac{v_2}{v_1}\right)^2 \cos^2(\theta_c)} \qquad (2.22)$$

Next, recalling the fact that $\mu_1 = \mu_2$ and $Z = (\mu/\varepsilon)^{1/2}$, it can be shown that*

$$\boxed{\theta_c = \sin^{-1}\left(\sqrt{\frac{\varepsilon_1}{\varepsilon_1 + \varepsilon_2}}\right)} \qquad (2.23)$$

* Here again the definition of critical angle here is relative to horizontal, so this expression differs from that given in Ref. 11 and other EM texts.

The critical angle exists for TM waves regardless of whether $\varepsilon_2/\varepsilon_1$ is greater than or less than one. Since this effect only occurs for TM waves, it can be used to separate out the TM component from a circularly (or randomly) polarized wave upon reflection. Thus this angle is also called the *polarizing angle* or sometimes the *Brewster angle*, particularly in radar work.

It is also interesting to note that in the case where $\varepsilon_2/\varepsilon_1 < 1$, it is possible for the transmission coefficient to equal zero. This occurs when Z_L goes to zero, or when

$$1 - \frac{\varepsilon_1}{\varepsilon_2}\cos^2(\theta) = 0$$

or

$$\theta = \cos^{-1}\left(\sqrt{\frac{\varepsilon_2}{\varepsilon_1}}\right) \tag{2.24}$$

This only occurs when the wave is traversing a boundary between dielectrics in the direction of the larger dielectric constant. This angle represents the point where there is no transmission and the wave is totally reflected.

2.6.2 Dielectric-to-Perfect Conductor Boundaries

For a wave in a dielectric that is incident on a perfect conductor, the resulting reflection and transmission coefficients are

$$\text{TM:} \quad \rho = -1$$
$$\text{TE:} \quad \rho = 1$$
$$\tau = 0$$

as expected. Thus a vertically polarized wave will undergo a 180-degree phase shift upon reflection, whereas a horizontally polarized wave does not experience a phase shift. Thus reflection of off perfectly conductive smooth ground has the interesting effect of converting right-hand circular polarization to left-hand circular polarization and vice versa.

2.6.3 Dielectric-to-Lossy Dielectric Boundary

When considering the effects of a lossy-dielectric boundary on an electromagnetic wave, the complex permittivity can be used in the prior expressions to determine the reflection and transmission coefficients. The resulting coefficients will be complex and thus will have both a magnitude and a phase. The resulting reflected and transmitted components of the wave will be at same angles as those predicted in previous sections, whereas the amplitudes and phases will be modified based on the material properties. In practical RF propagation problems, it is rare that the surface material is sufficiently smooth and the properties sufficiently understood to permit a detailed analysis of the surface interaction. Thus the predicted interaction is usually handled empirically. There are some exceptions, including RF modeling software, where assumptions are applied to the reflecting surface and numerical techniques used to accommodate the surface irregularities. Another case where an exact analysis may be practical is interaction with a pane of glass. If the material properties are known, then the analysis is straightforward.

2.7   PROPAGATION IMPAIRMENT

There are a variety of phenomena that occur when an electromagnetic wave is incident on a surface. These phenomena depend upon the polarization of the wave, the geometry of the surface, the material properties of the surface, and the characteristics of the surface relative to the wavelength of the electromagnetic wave.

Reflection: Whenever an electromagnetic wave is incident on a smooth surface (or certain sharp edges), a portion of the wave will be reflected. This reflection can be thought of as specular, where the grazing angle and reflection angle are equal.

Scattering: Scattering occurs when an electromagnetic wave is incident on a rough or irregular surface. When a wave is scattered, the resulting reflections occur in many different directions. When looked at on a small scale, the surface can often be analyzed as a collection of flat or sharp reflectors. The determination of when a surface is considered rough is usually based on the Rayleigh roughness criterion, which is discussed in Chapter 8.

Diffraction: Diffraction occurs when the path of an electromagnetic wave is blocked by an obstacle with a relatively sharp edge (as compared to the wavelength of the wave). The effect of diffraction is to fill in the shadow that is generated by the blockage.

Refraction: Is the alteration of the direction of travel of the wave as the transmitted portion enters the second material (i.e., penetrates the surface).

Absorption: Anytime that an electromagnetic wave is present in a material other than free space, there will be some loss of strength with distance due to ohmic losses.

Depolarization: As shown previously, the effects of transmission and reflection depend upon the orientation of the incident wave's polarization relative to the plane of incidence. This can have the effect of altering the polarization of the transmitted and reflected wave, particularly if the incident wave is circularly or elliptically polarized.

2.8 GROUND EFFECTS ON CIRCULAR POLARIZATION

Since it is possible for either diffraction or ground reflection to affect vertical and horizontal polarization differently, it is worthwhile to examine the impact such impairments have on a circularly polarized signal.

A circularly polarized wave (RHCP or LHCP) can be represented as the sum of two orthogonal linearly polarized waves. The polarization vectors are orthogonal to each other, and the respective signals are also orthogonal in phase. For ideal circular polarization, the components will be of equal amplitude and 90 degrees out of phase. For real-world antennas, however, the polarization is actually elliptical (circular is a special case of elliptical polarization). Thus the two vectors are not of equal amplitude. The *axial ratio* is defined as the voltage ratio of the major axis to the minor axis of the polorization ellipse (Chapter 3). The axial ratio of the transmitted wave is given by

$$AR_t = \frac{|E_{\text{maj}}|}{|E_{\text{min}}|} \tag{2.25}$$

The transmit antenna axial ratio is a specified parameter of the transmit antenna, nominally 1–3 dB.

The transmitted wave can be represented as

$$E_t = \left(\frac{E}{\sqrt{AR_t}} + j\sqrt{AR_t}\,E \right)\sqrt{\frac{AR_t}{1 + AR_t^2}}$$

where the (normalized) transmitted power is

$$P_t = \frac{1}{2}E^2$$

and the resulting axial ratio is AR_t.

Let the path gain (i.e., total path loss) for the horizontal and vertical components be designated as G_H and G_V, respectively. Since the path introduces loss (reduction in amplitude), each of these gains will be a very small number.

Since the orientation of the transmit antenna polarization ellipse is unknown, it is assumed that the major axis is parallel to whichever linear polarization experiences the greatest gain (least loss) to get a worst-case assessment. Thus the power gains applied to each component of the transmitted wave are

$$G_A = \max(G_H, G_V)$$
$$G_B = \min(G_H, G_V)$$

$$E_w = \left(\frac{E}{\sqrt{AR_t}} \sqrt{G_B} + j\sqrt{AR_t} E \sqrt{G_A} \right) \sqrt{\frac{AR_t}{1 + AR_t^2}} \qquad (2.26)$$

The axial ratio of the received wave is

$$AR_w = AR_t \frac{\sqrt{G_A}}{\sqrt{G_B}} \qquad (2.27)$$

The *polarization loss factor* at the receive antenna is given by [12]

$$F = \frac{(1 + AR_w^2)(1 + AR_r^2) + 4 AR_w AR_r + (1 - AR_w^2)(1 - AR_r^2)\cos(2\theta)}{2(1 + AR_w^2)(1 + AR_r^2)} \qquad (2.28)$$

where θ is the difference between the polarization tilt angles, taken to be 90 degrees for the worst case. Making the apparent substitutions, the expression for the polarization loss factor becomes

$$F = \frac{\left(1 + AR_t^2 \dfrac{G_A}{G_B}\right)(1 + AR_r^2) + 4 AR_t \sqrt{\dfrac{G_A}{G_B}} AR_r - \left(1 - AR_t^2 \dfrac{G_A}{G_B}\right)(1 - AR_r^2)}{2\left(1 + AR_t^2 \dfrac{G_A}{G_B}\right)(1 + AR_r^2)}$$

$$(2.29)$$

The transmit and receive antennas are identical, so

$$AR = AR_t = AR_r$$

and

$$\boxed{F = \frac{\left(1 + AR^2 \dfrac{G_A}{G_B}\right)(1 + AR^2) + 4 AR^2 \sqrt{\dfrac{G_A}{G_B}} - \left(1 - AR^2 \dfrac{G_A}{G_B}\right)(1 - AR^2)}{2\left(1 + AR^2 \dfrac{G_A}{G_B}\right)(1 + AR^2)}} \qquad (2.30)$$

The overall loss that is applied to the transmitted EIRP to determine the received power level is

$$\boxed{L = F G_A} \qquad (2.31)$$

To verify this expression, setting the antenna axial ratios equal to one and taking the limit as G_B goes to zero yields $L = G_A/2$. This is the expected result, analogous to transmitting perfect linear polarization and receiving with a perfect circularly polarized antenna.

2.9   SUMMARY

In this chapter the basics of electromagnetic theory were briefly reviewed, followed by a discussion of electromagnetic waves and how they behave in different materials. Of particular interest in RF propagation is the interaction of an electromagnetic wave with a boundary between two dielectrics or between a dielectric and a conductor. Using an analysis similar to transmission line theory, expressions for the reflection coefficient, transmission coefficient, and transmission angle were developed. The polarization of an electromagnetic wave is defined as the orientation of the plane that contains the electric field component of the wave. The polarizations of most interest are linear: (usually vertical or horizontal) and elliptical or circular polarization, which can be expressed as a linear combination of vertically and horizontally polarized waves with appropriate phase.

The interaction of electromagnetic waves with a material boundary is a strong function of the orientation of the polarization vector relative to the plane of incidence. Two key cases are defined: transverse magnetic or TM, where the electric field is contained in the plane of incidence and the magnetic field is transverse to the plane of incidence and transverse electric or TE, where the magnetic field is contained in the plane of incidence and the electric field is transverse to the plane of incidence. These two polarizations are analogous to vertically and horizontally polarized wave intersecting a horizontal boundary (such as the earth's surface).

The effect of a material boundary on an electromagnetic wave is a function not only of the orientation of the polarization vector of the wave, but also of the angle of incidence (grazing angle) and the properties of both materials. In the case of two dielectrics, the dielectric constants determine to a large extent the effect that the boundary will have on the wave. When a wave is incident from a dielectric to a conductor, the wave will only slightly penetrate the conductor and the majority of the wave will be reflected.

Due to the dependence of transmission and reflection on the polarization of the wave, a circularly polarized wave will, in general, become depolarized when reflected off of a material boundary. The effect of terrestrial interactions on an elliptically polarized wave is examined and an expression for the resulting polarization loss, taking into account the axial ratio of the transmitting and receiving antennas, is developed.

REFERENCES

1. M. A. Plonus, *Applied Electromagnetics*, McGraw-Hill, New York, 1978, p. 124.
2. M. A. Plonus, *Applied Electromagnetics*, McGraw-Hill, New York, 1978, Table 1.1, p. 3.
3. M. A. Plonus, *Applied Electromagnetics*, McGraw-Hill, New York, 1978, pp. 32–35.
4. S. V. Marshall and G. G. Skitek, *Electromagnetic Concepts & Applications*, 2nd ed., Prentice-Hall, Englewood Cliffs, NJ, 1987, pp. 141–143.
5. M. A. Plonus, *Applied Electromagnetics*, McGraw-Hill, New York, 1978, Table 2.1, p. 62.
6. M. A. Plonus, *Applied Electromagnetics*, McGraw-Hill, New York, 1978, Chapter 6.
7. S. Ramo, J. R. Whinnery, and T. Van Duzer, *Fields and Waves in Communication Electronics*, 2nd ed., John Wiley & Sons, New York, 1984, p. 131.
8. S. Ramo, J. R. Whinnery, and T. Van Duzer, *Fields and Waves in Communication Electronics*, 2nd ed., John Wiley & Sons, New York, 1984, p. 280.
9. S. Ramo, J. R. Whinnery, and T. Van Duzer, *Fields and Waves in Communication Electronics*, 2nd ed., John Wiley & Sons, New York, 1984, Chapter 6.
10. S. Ramo, J. R. Whinnery, and T. Van Duzer, *Fields and Waves in Communication Electronics*, 2nd ed., John Wiley & Sons, New York, 1984, pp. 302–303.
11. S. Ramo, J. R. Whinnery, and T. Van Duzer, *Fields and Waves in Communication Electronics*, 2nd ed., John Wiley & Sons, New York, 1984, p. 308.
12. J. S. Hollis, T. J. Lyon, and L. Clayton, *Microwave Antenna Measurements*, 2nd ed., Scientific Atlanta, Atlanta, GA, 1970, Chapter 3.

EXERCISES

1. Find the electric field vector (direction and intensity) for the following;
   (a) At a distance of 1 m from a 5-C point charge, if the medium is polystyrene.
   (b) At a distance of 0.5 m from an infinite line charge of 2 C, if the medium is distilled water.
   (c) At a distance of 2 m from an infinite plane charge of 1 C in air.
   (d) At a distance of 20 m from an infinite plane charge of 1 C in air.

2. Determine the angle of refraction of an electric field that that is incident on a body of calm water at an angle of 30 degrees relative to the horizon. The electric field originates in air, and the water may be considered pure and treated as distilled water.

3. Find the magnetic field vector at a distance of 1 m from a conductor carrying a 1-A current. The conductor is oriented vertically in free space, with the current flowing in the positive z direction.

4. Consider a vertically polarized electromagnetic wave in air, with a frequency of 10 MHz. If the wave is incident on a flat, horizontal surface

of polystyrene, find the magnitude and angle of both the reflected and refracted portion of the wave?

5. Repeat problem 4 if the wave is horizontally polarized.

6. Consider a 100-MHz wave in free space.
   (a) If the wave is incident on a flat copper sheet, what is the skin depth?
   (b) If the copper sheet is to be used for shielding and 97% of the electromagnetic wave must be blocked by the shield, how thick should the copper sheet be?

7. What is the worst-case polarization loss factor if a wave is transmitted from an antenna with a 2-dB axial ratio and received by an antenna with a 2.5-dB axial ratio? Assume that there is no distortion introduced by the medium between the two antennas.

CHAPTER 3

Antenna Fundamentals

3.1 INTRODUCTION

Every wireless system must employ an antenna to radiate and receive electromagnetic energy. The antenna is the transducer between the system and free space and is sometimes referred to as the air interface. While a comprehensive treatment of the subject of antennas is beyond the scope of this work, it is helpful to develop an intuitive understanding about how they operate and provide a few examples of different antenna architectures. There are some useful rules of thumb for relating antenna parameters to antenna size and shape. For many types of antennas it is possible to estimate the gain from the physical dimensions or from knowledge of the antenna beamwidths.

A fundamental principle of antennas, called *reciprocity*, states that antenna performance is the same whether radiation or reception is considered [1]. The implication of this principle is that antenna parameters may be measured either transmitting or receiving. The principle of reciprocity means that estimates of antenna gain, beamwidth, and polarization are the same for both transmit and receive. This principle is used freely in this text as well as in most works on antennas.

3.2 ANTENNA PARAMETERS

A useful abstraction in the study of antennas is the *isotropic radiator*, which is an ideal antenna that radiates (or receives) equally in all directions, with a spherical pattern. The isotropic radiator is also sometimes called an omnidirectional antenna, but this term is usually reserved for an antenna that radiates equally in all directions in one plane, such as a whip antenna, which radiates equally over azimuth angles but varies with elevation. The power density, S, due to an isotropic radiator is a function only of the distance, d, from the antenna and can be expressed as the total power divided by the area of a sphere with radius d.

Introduction to RF Propagation, by John S. Seybold

$$S = \frac{P}{4\pi d^2} \quad \text{W/m}^2 \tag{3.1}$$

That is, the power is uniformly distributed over the sphere. Thus for an isotropic radiator, the power density at a given range is constant over all angles and is equal to the average power density at that range.

3.2.1 Gain

For a real antenna, there will be certain angles of radiation, which provide greater power density than others (when measured at the same range). The *directivity* of an antenna is defined as the ratio of the radiated power density at distance, d, in the direction of maximum intensity to the average power density over all angles at distance, d. This is equivalent to the ratio of the peak power density at distance d, to the average power density at d:

$$D = \frac{\text{Power density at } d \text{ in direction of maximum power}}{\text{Mean power density at } d}$$

Thus an isotropic antenna has a directivity of $D = 1$. When the antenna losses are included in the directivity, this becomes the antenna gain

$$G = \eta \cdot \frac{\text{Power density at } d \text{ in max direction}}{P_T / 4\pi d^2}$$

where

P_T is the power applied to the antenna terminals
$4\pi d^2$ is the surface area of a sphere with radius d
η is the total antenna efficiency, which accounts for all losses in the antenna, including resistive and taper* losses ($\eta = \eta_T \eta_R$)

Antenna gain can be described as the power output, in a particular direction, compared to that produced in any direction by an isotropic radiator. The gain is usually expressed in dBi, decibels relative to an ideal isotropic radiator.

3.2.2 Effective Area

An *aperture antenna* is one that uses a two-dimensional aperture such as a horn or a parabolic reflector (as opposed to a wire antenna). The gain of an

---

* Taper loss is the inefficiency that is introduced when nonuniform illumination of the aperture is used. Nonuniform aperture illumination is used to control radiation pattern sidelobes, similar to windowing in spectral analysis and digital filtering [2, 3].

aperture antenna, such as a parabolic reflector, can be computed using an *effective area*, or *capture area*, which is defined as

$$A_e = \eta A_p \quad m^2 \tag{3.2}$$

where A_p is the physical area of the antenna and η is the overall efficiency of the antenna (generally ranging from 50% to 80%). The expression for the gain of an aperture antenna is

$$\boxed{G = \frac{4\pi A_e}{\lambda^2}} \tag{3.3}$$

If only the physical dimensions of the antenna are known, one may assume an efficiency of about 0.6 and estimate the gain reasonably well for most antennas using the above expression.

Example 3.1. Given a circular aperture antenna with a 30-cm diameter, what is the antenna gain at 39 GHz?

A 30-cm-diameter circular aperture has a physical area of $0.0707\,m^2$. Assuming an aperture efficiency of 60% yields

$$A_e = 0.0424\ m^2$$

Using the expression for antenna gain and the wavelength of 7.69 mm yields

$$G = 9005$$

which, expressed as decibels relative to an isotropic radiator, is $10\log(9005)$, or 39.5 dBi of gain. □

If the physical dimensions of an aperture antenna are not known, but the azimuth and elevation 3-dB beamwidths, θ_{AZ} and θ_{EL}, are known, the gain can be estimated using the following rule of thumb:

$$G \cong \frac{26,000}{\theta_{AZ}\theta_{EL}}$$

where the 3-dB beamwidths are expressed in degrees [4].

The effective or capture area of a linear antenna, such as a vertical whip, a dipole, or a beam, does not have the same physical significance as it does for an aperture antenna. For example, a center-fed, half-wave dipole has an effective capture area of $0.119\lambda^2$ [5]. Instead of effective area, the concept of *effective height* is employed for linear antennas [6]. The effective height of an antenna is the height that, when multiplied by the incident electric field, results in the voltage that would actually be measured from the antenna.

$$V = h_e E \qquad \text{volts}$$

or

$$h_e = V/E \qquad \text{m}$$

where V is the voltage measured at the antenna terminals and E is the magnitude of the incident electric field.

Another way to view the effective height is by considering the average of the normalized current distribution over the length of the antenna and multiplying that value by the physical length of the antenna.

$$h_e = \frac{1}{I_0} \int_0^l I(x)\, dx \qquad (3.4)$$

Using this information, the effective area of a linear antenna can be found. The current distribution over a half-wave dipole is nearly sinusoidal [7]

$$I(x) = I_0 \sin(\pi x)$$

and for a very short dipole ($l < 0.1\lambda$) it is approximately triangular, with the peak current occurring at the center and zero current at each end.

The relationship between the effective height and the effective area is given by

$$A_e = \frac{h_e^2 Z_o}{4R_r} \qquad (3.5)$$

where

A_e is the effective area in square meters
h_e is the effective height in meters
Z_o is the intrinsic impedance of free space, 377 ohms
R_r is the radiation resistance of the antenna in ohms

Example 3.2. Determine the effective height and the effective area of a half-wave dipole at 500 MHz.

A half-wave dipole for 500 MHz has a physical length of

$$l = \frac{1}{2} \frac{c}{500\ \text{MHz}} = 0.3\ \text{m}$$

For a half-wave dipole, the current distribution is

$$I(x) = I_0 \sin(\pi x)$$

which yields a normalized average current value of $(2/\pi)I_0$ or $0.637I_0$. The effective height is then found from (3.4):

$$h_e = 0.191 \text{ m}$$

Using the radiation resistance for a half-wave dipole, 73 ohms [8], and the impedance of free space, one can readily determine the effective area using (3.5):

$$A_e = 0.047 \text{ m}^2$$

Note that this effective area has no direct relationship to the physical cross section of the dipole antenna. That is to say, it does not depend upon the thickness of the dipole, only its length. □

3.2.3 Radiation Pattern

The radiation pattern of an antenna is a graphical depiction of the gain of the antenna (usually expressed in dB) versus angle. Precisely speaking, this will be a two-dimensional pattern, a function of both the azimuth and elevation angles. This is illustrated in Figure 3.1 for a circular aperture antenna. In most

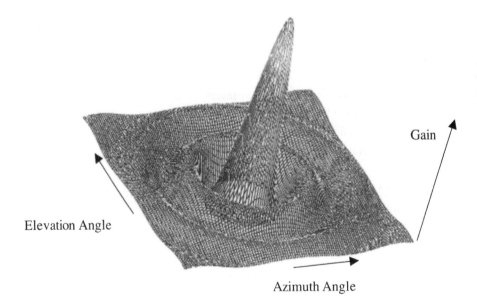

Figure 3.1   Three-dimensional plot of a two-dimensional antenna radiation pattern.

cases, however, looking at the principal plane cuts—that is, an azimuth pattern while bore-sighted in elevation and vice versa—is sufficient. Figure 3.2 shows a typical principal plane pattern. Antenna patterns always describe the far-field pattern, where the gain or directivity is a function of the azimuth and elevation angles only and is independent of distance, d.

The mainlobe of the antenna is the lobe where the peak gain occurs. The beamwidth of an antenna is defined as the angular distance between the two points on the antenna pattern mainlobe that are 3 dB below the maximum gain point. From Figure 3.2, the 3-dB beamwidth of that antenna may be estimated to be about 5 degrees (±2.5). Another parameter that is often of considerable interest is the maximum or peak sidelobe level. The pattern in Figure 3.2 shows a maximum sidelobe level of about −17 dB relative to the beam peak, which has been normalized to 0-dB gain. Low sidelobes reduce the risk of undesired signal radiation or reception, which can be valuable in high traffic or multipath environments. Sidelobe reduction is obtained, however, at the expense of widening the mainlobe and reducing the antenna gain, or by increasing the antenna size.

Another often-quoted specification of directional antennas is the front-to-back ratio. This is the ratio of the antenna gain at 0 and 180 degrees azimuth and provides an indication of how well the antenna will reject interfering signals that arrive from the rear of the antenna. The front-to-back ratio is a very important parameter when planning frequency reuse and interference reduction.

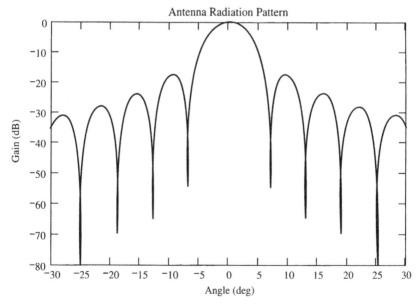

Figure 3.2 Typical antenna radiation pattern (normalized gain) in one dimension.

The antenna pattern of an aperture antenna is a function of the illumination taper across the aperture. The relationship between the spatial energy distribution across the aperture and the gain pattern versus angle is an inverse Fourier transform [9]. Thus a uniformly illuminated aperture will produce the narrowest main beam and highest possible gain at the expense of producing sidelobes that are only 13 dB below the peak. By tapering the illumination using a window function, the gain is slightly reduced (taper loss) and the mainlobe is broadened, while the sidelobes are reduced in amplitude. Illumination taper functions such as a raised cosine are often used to produce antennas with acceptable sidelobes. Whenever an illumination taper is used, the resulting antenna efficiency (and therefore gain) is reduced.

3.2.4 Polarization

Polarization is defined as the orientation of the plane that contains the electric field component of the radiated waveform. In many cases, the polarization of an antenna can be determined by inspection. For instance, a vertical whip antenna generates and receives vertical polarization. Similarly, if the antenna element is horizontal, the wave polarization will be horizontal. Vertical and horizontal polarizations are both considered linear polarizations. Another type of polarization is circular or elliptical polarization. Circular polarization is similar to linear polarization, except that the polarization vector rotates either clockwise or counterclockwise, producing right-hand circular or left-hand circular polarization. Circular polarization is a special case of elliptical polarization, where the vertical and horizontal components of the polarization vector are of equal magnitude. In general, aperture antennas can support vertical, horizontal, or elliptical polarization, depending upon the type of feed that is used.

3.2.5 Impedance and VSWR

An antenna presents a load impedance or driving point impedance to whatever system is connected to its terminals. The driving point impedance is ideally equal to the radiation resistance of the antenna. In practical antennas, the driving point impedance will also include resistive losses within the antenna and other complex impedance contributors such as cabling and connectors within the antenna. The driving point impedance of an antenna is important in that a good impedance match between the circuit (such as a transceiver) and the antenna is required for maximum power transfer. Maximum power transfer occurs when the circuit and antenna impedances are matched.* Maximum power transfer is desirable for both transmitting and receiving.

---

* Actually, maximum power transfer required a complex conjugate match. Most antenna systems have real or nearly real driving point impedance, in which case a conjugate match and a match are identical. An antenna connected to a cable with a characteristic impedance that is different from its driving point impedance will present a complex impedance at the other end of the cable. In such cases, a transmatch can be used to perform the conjugate matching.

When the antenna and circuit impedances are not matched, the result is reduced antenna efficiency because part of the signal is reflected back to the source. The square root of ratio of the reflected power to the incident power is called the *reflection coefficient*.

$$|\rho| = \sqrt{\frac{P_r}{P_i}}$$

Clearly, the reflection coefficient must be between zero and one (inclusive). The reflection coefficient can be determined from the circuit and antenna impedances,

$$\rho = \frac{Z_1 - Z_0}{Z_1 + Z_0} \tag{3.6}$$

The amount of signal passing between the transceiver and the antenna is

$$P_t = (1 - \rho^2) P_i$$

Thus the impedance mismatch loss is [10]

$$L_m = 1 - \rho^2 \tag{3.7}$$

If there is a cable between the antenna and the transceiver, the mismatch creates a voltage standing wave ratio (VSWR, or often just SWR) on the cable. The effect of a VSWR on a cable is to increase the effect of the cable loss [11]. One way to compute the VSWR is

$$\text{VSWR} = \frac{1 + |\rho|}{1 - |\rho|} \tag{3.8}$$

This can also be expressed in terms of the of antenna and transceiver characteristic impedances.

Example 3.3. What is the reflection coefficient, VSWR and matching loss for a 50-ohm transceiver driving a half-wave dipole?

The half-wave dipole has a characteristic impedance of 73 ohms, so the reflection coefficient is ±0.23, depending upon the assignment of Z_1 and Z_0. The corresponding VSWR from (3.8) is 1.6. Applying (3.7), the matching loss is found to be 0.947, or −0.24 dB, which is negligible in most applications. □

3.3 ANTENNA RADIATION REGIONS

The antenna radiation field is divided into three distinct regions, where the characteristics of the radiated wave are different. While these are not firm boundaries, they represent convention usage and provide some insight to the actual radiated field as a function of distance from the antenna.

The *far-field* (Fraunhoffer) region is defined for distances such that

$$d > \frac{2D^2}{\lambda} \tag{3.9}$$

where D is the largest linear dimension of the antenna, and λ is the wavelength. This is the region where the wavefront becomes approximately planar (less than 22.5 degrees of phase curvature over a similar-sized aperture) [12]. In the far-field region, the gain of the antenna is a function only of angle (i.e., the antenna pattern is completely formed and does not vary with distance). In the far field, the electric and magnetic field vectors are orthogonal to each other. For electrically small antennas ($D \ll \lambda$) the far-field boundary defined by the preceding equation may actually fall within the near-field region. In such cases, the far-field boundary should be taken as equal to the near-field boundary rather than within it.

The *radiating near-field* is the region between the far-field region and the reactive near-field region, also called the *transition region*.

$$\frac{\lambda}{2\pi} < r < \frac{2D^2}{\lambda} \tag{3.10}$$

In this region, the antenna pattern is taking shape but is not fully formed. Thus the antenna gain will vary with distance even at a fixed angle. The radiated wave front is still clearly curved (nonplanar) in this region and the electrical and magnetic field vectors are not orthogonal.

The *reactive near-field* is defined as

$$r < \frac{\lambda}{2\pi} \tag{3.11}$$

This region is measured from the phase center or center of radiation of the antenna and will be very close to the surface of the antenna. In general, objects within this region will result in coupling with the antenna and distortion of the ultimate far-field antenna pattern. Figure 3.3 is an illustration of the radiation regions for a reflector antenna.

Analysis of near-field coupling can be deceiving. It is unlikely that anything will be close enough to a 40-GHz antenna to cause near-field coupling since the wavelength is less than a centimeter, making the reactive near-field boundary on the order of a couple of millimeters. On the other hand, it is very difficult to avoid coupling with an HF antenna for amateur radio. Antennas for the 75-m amateur radio band (4MHz) have a reactive near-field boundary of 12m. Thus any large conductor (relative to the wavelength) within about

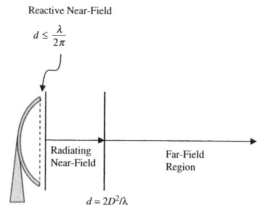

Figure 3.3 Typical boundaries for antenna radiation regions.

39 ft of the antenna will couple with the antenna and "detune" it. The result can be an altered resonant frequency, radiation resistance (defined shortly), and/or radiation pattern. This can be a significant concern for radio amateurs operating from small lots, where overhead utility lines, house gutters, fascia trim, and other such items are difficult to avoid. Tying these conductors to a good earth ground eliminates their resonant effect and mitigates their impact.

Example 3.4. How much separation is required between a 140-MHz quarter-wave monopole antenna and a 450-MHz quarter-wave monopole antenna if both are mounted on the roof of an automobile and near-field coupling is to be avoided?

To avoid mutual, near-field coupling, each antenna must be outside of the reactive near-field of the other. The reactive near-field of the 140-MHz antenna is larger than that of the 440-MHz antenna, so it determines the minimum required separation.

$$d_{min} \leq \frac{\lambda}{2\pi}$$

where

$$\lambda = 2.14 \text{ m}$$

The result is that the minimum required separation is

$$d_{min} = 0.341 \text{ m} \quad \square$$

3.4 SOME COMMON ANTENNAS

The possible configurations of an antenna are limited only by the imagination. In the field of electromagnetic compatibility, any conductor is a potential radiator and may be treated as an antenna. In this section we examine several broad categories of widely used antennas. The reader is directed to the references for more detailed information on these antennas or for insight into other antenna configurations.

3.4.1 The Dipole

The half-wave dipole is configured as shown in Figure 3.4. It usually consists of two segments, each one-quarter of a wavelength long, with the feed in the middle, although offset-feed and end-fed dipoles are also used. The rings in Figure 3.4 illustrate the directionality of the radiation pattern of the dipole. It can be seen that there are areas of reduced gain off the ends of the elements. The gain of a half-wave dipole is theoretically 2.14 dB. In general, antenna gain may be specified in decibels relative to an isotropic radiator, dBi, or in decibels relative to an ideal half-wave dipole, dBd.

The quarter-wave vertical or monopole antenna is one-quarter of a wavelength long and is often (or at least ideally) mounted on an infinite, perfectly reflective (conductive) ground plane. The ground plane provides a reflection of the antenna, causing it to behave as a dipole antenna. Oftentimes in practical use, the "ground plane" turns out to be more of a lossy counterpoise (such as the transceiver case and the hand and arm holding it), rather than a ground plane that can provide an electrical image of the antenna. Even then, the

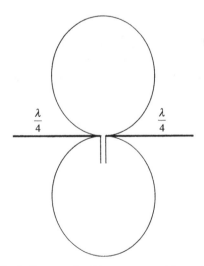

Figure 3.4   The half-wave dipole antenna and its radiation pattern.

quarter-wave vertical antenna is still relatively effective. The counterpoise can be thought of as a nonideal implementation of the other half of a half-wave dipole.

The quarter-wave antenna may also be physically shortened as long as the electrical length of $\lambda/4$ is maintained. An example of such an antenna is the flexible rubber duck or helix antenna [13] used on many handheld VHF and UHF radios. While convenient, these antennas can have considerably less gain than a regular quarter-wave antenna, on the order of −3 dB or more. All quarter-wave antennas perform best when they have a large conductive ground plane. Using the human hand and arm as a lossy counterpoise degrades both the radiation pattern and the gain. If the antenna is held near the body such as next to the head or on the waist, the net gain may be further reduced by 10 dB or more.

An important parameter in antenna performance is the radiation resistance. The radiation resistance of a resonant (half-wave) dipole is approximately 73 ohms, if it is center-fed. For an ideal antenna (no resistive losses), the driving-point impedance seen by the transmitter or receiver is equal to the radiation resistance. The same dipole can be end-fed, but the driving-point impedance then becomes extremely large, on the order of 3000 ohms. A matching network of some type is then required to match the radiation resistance to the feedline and receiver/transmitter impedance to avoid large reflections and SWR (standing-wave ratio, also called voltage standing wave ratio or VSWR). Such matching networks (transmatch) can be somewhat lossy and reduce the efficiency of the antenna.

It is not unusual to see a "loading coil" at the base of an antenna, particularly for mobile applications. This loading coil can be either an impedance matching device if the radiator is an end-fed dipole, or it may be an inductor that is used to increase the electrical length of the antenna. An example is the "Hamstick" antenna used for HF work in automobiles. The antenna is a 4-ft piece of fiberglass with windings running the entire length, forming what is called a normal mode helix. At the top is a 3- to 4-ft whip that is adjusted in length to tune the resonant frequency of the antenna. The body of the car serves as a counterpoise, and the 8-ft antenna is then resonant at an HF frequency such as 3.9 MHz, giving it a resonant length of 75 m. This is shown conceptually in Figure 3.5. Though the antenna is resonant, it does not have the same gain as a full-length dipole and in fact its gain considerably less than unity.

It is also possible to locate "traps" (tank or resonant circuits) along the length of a radiator, so that the same vertical antenna can be resonant at several spot frequencies. Such traps are resonant lumped elements. A trap alters the electrical length of the antenna in one of two ways. It may be used to cut off frequencies above its resonant frequency, so that the remainder of the antenna is invisible above the resonant frequency. The other approach is to have the trap resonance set in between two frequencies of interest, so that at the higher frequency it becomes a capacitive load, effectively shortening the

Figure 3.5 Base-loaded mobile HF antenna.

electrical length, and at the lower frequency it becomes an inductive load, thereby increasing the electrical length. Another architecture for the multiband vertical antenna is a parallel array of closely spaced vertical radiators, each tuned to a different band. The close proximity of the radiators means that they have strong interaction, which alters the electrical length of the radiators, so they do not physically measure a quarter- or a half-wavelength. The parallel radiators are functioning as distributed elements. Both of these antenna architectures are illustrated in Figure 3.6.

3.4.2 Beam Antennas

The half-wave dipole antenna also serves as a component of another popular antenna, the Yagi-Uda (Yagi) beam antenna. The Yagi antenna is comprised of a driven element, which is a center-fed half-wave dipole, a reflector element, which is slightly longer than the driven element, and (optionally) several director elements that are progressively smaller than the driven element. This is shown in Figure 3.7. The boom may be conductive or nonconductive, with the element lengths and antenna properties being slightly different in each case. The Yagi beam is a directional antenna that generates a linearly polarized wave. The gain and front-to-back ratio can be substantial if a long boom and a sufficient number of elements are used in the design. Section 5.4 in Ref. 14 provides an excellent treatment of Yagi antenna design and analysis.

When mounted with the elements oriented vertically, the polarization will be vertical and vice versa. Conventional television broadcast signals are horizontally polarized, which is why television beam antennas are mounted with the elements oriented horizontally. Horizontal polarization was chosen for television broadcast because man-made interference tends to be vertically polarized. By using horizontal polarization, television receivers have less interference to contend with. Mobile radio uses vertical polarization primarily for convenience because it is difficult to mount a horizontally polarized

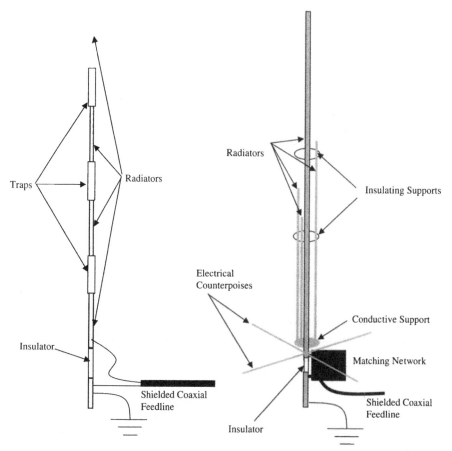

Figure 3.6   Four-band quarter-wave trap vertical and a four-band half-wave parallel radiator vertical antenna.

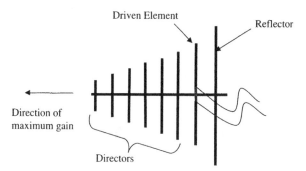

Figure 3.7   A typical Yagi-Uda antenna (the taper in the element lengths has been exaggerated in this picture for clarity).

antenna on a automobile or a handset, but also because of its reflective properties.

3.4.3   Horn Antennas

A horn antenna can be thought of as a flared end of a waveguide. A horn antenna is one example of an aperture antenna. The polarization of the emitted signal is dependent upon the polarization of the waveguide transducer. There are various shapes of horn antennas, and the taper of the flare and the aperture size control the resulting radiation pattern. Horn antennas can provide very high gain and narrow beams, depending upon their design. They are frequently used as gain or calibration standards since their gains are very predictable and repeateable.

3.4.4   Reflector Antennas

Reflector antennas include the parabolic reflector or dish antenna and the Cassegrain antenna. The parabolic reflector has a receiver/transmitter mounted in a small unit that is usually located at the focal point of the parabolic dish. In addition to the efficiency loss due to illumination taper, the supports for the antenna feed produce blockage in the antenna field of view. This causes shadowing and produces diffraction lobes in the antenna pattern. An offset feed can be used to reduce the shadowing effect of the feed supports. An example of a parabolic dish antenna with an offset feed is the direct broadcast TV antenna as shown in Figure 3.8. The antenna feed (actually it is only a receive antenna) contains a low-noise amplifier and a broadband (block) down-converter that shifts all of the signals to L-band. The feed unit is therefore referred to as a low-noise block (LNB). The block of L-band signals then travels through the coaxial cable to the receiver, which selects the desired frequency for demodulation.

It is interesting that Direct TV broadcasts different programming simultaneously on right and left circular polarizations. The Direct TV receiver sends

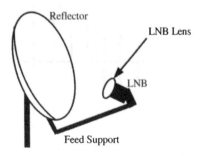

Figure 3.8   Offset feed reflector antenna for direct broadcast satellite TV.

a DC voltage over the receive signal coax to power the electronics in the LNB. The level of the DC also tells the block converter which polarization to receive based on the operator's channel selection at the receiver. This is why Direct TV receivers cannot directly share an antenna connection. For two receivers, a dual LNB is employed, with a separate cable running to each receiver. For more than two receivers, a dual LNB and a receiver control unit (multiswitch) are used to provide the appropriately polarized signal to each receiver, depending upon the channel being viewed. This is an example of frequency sharing by using polarization diversity.

For some applications, the receiver/transmitter unit is mounted behind the antenna with waveguide or coax cable connecting it to the center-mounted feed. Of course, this configuration entails cable or waveguide loss as the signal is fed to and from the feed point, so it is not suitable for most higher-frequency applications. One way to eliminate this additional loss is to employ a Cassegrain antenna, which is shown diagrammatically in Figure 3.9. The Cassegrain antenna also employs a parabolic reflector, but it has the antenna feed mounted in the center of the dish, where the signal radiates out to a sub-reflector that is mounted at the focal point of the parabolic reflector. The signal from the reflector then illuminates the dish and produces the radiation pattern. One advantage of the Cassegrain antenna over the basic parabolic reflector is that the receiver/transmitter unit can be mounted on the antenna without using cables or waveguides and are not required to be as small as those for a standard reflector antenna. Another advantage for very sensitive systems used on earth stations for satellite communication is that the sidelobes from the feed see sky rather than earth (since the feed is pointed upward), which reduces the antenna temperature and thereby the receiver noise floor. The center reflector still requires support, so the Cassegrain antenna also experiences blockage and diffraction lobes both from the subreflector supports and

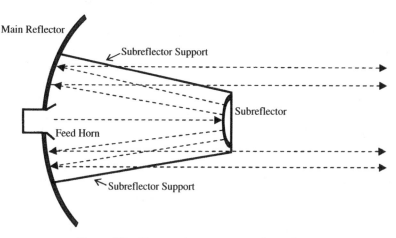

Figure 3.9 Cassegrain antenna configuration.

the subreflector itself. There are many different ways that the supports can be mounted, Figure 3.9 shows one possibility.

3.4.5   Phased Arrays

Phased array antennas were initially developed for large ground-based tracking radars. They provide the ability to focus a beam on each target in the field of view in rapid succession without requiring rapid repositioning of the aperture. The phased array is an array of radiating elements each with a controllable phase shifter. The antenna controller adjusts the phase (and sometimes also the gain) of each element to provide the desired beam position and beamwidth while controlling the sidelobes. Some systems even produce multiple beams simultaneously. Phased arrays suffer from a phenomenon called grating lobes when the look angle approaches 90 degrees off-boresight, which is why phased arrays sometimes include physical pointing. The beam-forming process uses digital signal processing techniques in the spatial domain to form beams and control sidelobes. Squinted beams can be formed to permit monopulse angle measurement.

As electronics have become increasingly powerful, efficient, and affordable, phased arrays have found application in many other areas. So-called smart antennas [15] have been used for GPS receivers and mobile telephone base stations for interference control. The advantage of the phased array is that it can steer a null in the direction of any detected interference, improving receiver operation. This is often accomplished using an adaptive algorithm to optimize signal-to-interference ratio.

3.4.6   Other Antennas

There are a multitude of antenna types, all of which cannot be even listed here. Some of the more noteworthy configurations are discussed. One interesting antenna is the magnetic field antenna. Most antennas are electric field antennas; they work by intercepting the electrical field component of the desired wave and converting it to a voltage that is sent to a receiver. It is equally possible to detect the magnetic field rather than the electric field, and it is sometimes advantageous to do so. One good example of a magnetic field antenna is the ferrite loop antenna in a small AM radio. The wavelength of AM broadcast is so long that electric field antennas are difficult to fabricate with any efficiency. Early AM radios required very long wire antennas. Later, as the receiver became more sensitive, moderately sized loop antennas where employed. Modern AM receivers use a ferrite rod with windings to sense the magnetic field of the AM wave and provide a signal to the receiver.

Lens antennas operate just as the name suggests. They are treated as an aperture antenna, but a lens covers the aperture and focuses the beam. The lenses are often made of plastic, but other materials can be used as long as they have the appropriate refractive properties at the frequency of interest.

Patch antennas are conformal elements, which are useful for applications where a low profile is required. An array of patch radiators can be used to form a beam. Such antennas are useful for mobile and particularly airborne applications, where low profile equates to less wind resistance and lower fuel costs.

3.5 ANTENNA POLARIZATION

Antenna polarization is defined as the polarization of the electromagnetic wave that it radiates. Polarization is defined in the far-field of the antenna and is the orientation of the plane that contains the electric field portion of the wave. In the far-field, the electrical field, magnetic field, and propagation direction vectors are all mutually orthogonal. This mutual orthogonality permits complete characterization of a wave by describing the electric field vector and the direction of propagation.

Most antennas and electromagnetic waves have either linear polarization (vertical or horizontal) or circular polarization, which can be right-hand or left-hand. In a general sense, all polarizations can be considered as special cases of elliptical polarization. This provides a convenient mathematical framework for looking at the effects of polarization. An elliptically polarized wave can be expressed as [16]

$$\mathbf{E} = \mathbf{x}E_x + \mathbf{y}E_y$$

where

$$E_x = E_1 \sin(\omega t - \beta z)$$

and

$$E_y = E_2 \sin(\omega t - \beta z + \delta)$$

are the **x** and **y** components of the **E** field vector. **E** is the vector that describes the electric field as a function of time and position along the **z** axis (direction of propagation). E_1 and E_2 are the amplitude of the linearly polarized wave in the **x** and **y** directions. The δ term describes the relative phase between the **x** and **y** components of the electric field vector. If δ is equal to zero, **E** will be linearly polarized, with the orientation of its electrical field determined by $\mathbf{x}E_1 + \mathbf{y}E_2$. If E_1 and E_2 are of equal magnitude and $\delta = \pm 90$ degrees, then the wave is circularly polarized. Circular polarizations, and elliptical polarizations in general, are defined as follows (Figure 3.10):

Right-Hand Circular Polarization. Clockwise rotation when viewed in the direction of propagation—that is, from behind (**z** into page).

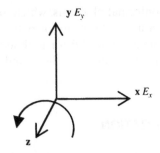

Figure 3.10 Relationship between orthogonal components of the **E** field and the direction of propagation for right circular polarization ($\delta = -90$ degrees).

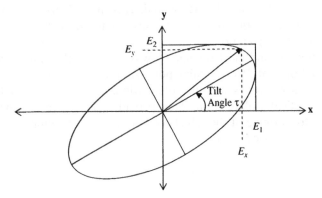

Figure 3.11 The polarization ellipse (Figure 2.22 from Ref. 16, courtesy of McGraw-Hill.)

> *Left-Hand Circular Polarization.* Counterclockwise rotation when viewed from behind (**z** into page).

These definitions correspond to the IEEE definitions [17, 18]. The labels are frequently shortened to right-circular and left-circular polarization. Using the definitions and the preceding equation, it is straightforward to verify that

$$\delta = +90° \rightarrow \text{Left circular polarization (\textbf{y} leads \textbf{x} by } 90°)$$

$$\delta = -90° \rightarrow \text{Right circular polarization}$$

Figure 3.11 shows the polarization ellipse as presented by Ref. 16. Note that E_1 and E_2 are the peak amplitudes in the **x** and **y** direction and are therefore fixed, whereas E_x and E_y are the instantaneous amplitudes of the **x** and **y** components, which are functions of time, t, position, z, and the relative phase, δ. This diagram makes it easy to visualize the following: If E_2 is zero (or if δ is zero), the polarization is linear horizontal; and if E_1 is zero, it is vertical polarization. The tilt angle, τ, is the orientation of the major axis of the polarization ellipse relative to the **x** axis.

The *axial ratio* is an important antenna parameter that describes the shape of the polarization ellipse. The axial ratio is defined as the ratio of the major axis to minor axis of the polarization ellipse when the phase angle between the linear polarization components, δ, is ± 90 degrees. The axial ratio is always greater than or equal to one. Since it is a ratio of amplitudes, it can be expressed in dB using $20\log(AR)$. For precise applications such as satellite communication, the axial ratio is given for circularly polarized antennas, as a metric of the antenna's deviation from ideal circular polarization. This is valuable for determining the potential cross-coupling between an incident wave and the orthogonal polarization channel of the receive antenna and the coupling loss between an incident wave and the receive antenna at the same polarization. Strictly speaking, if the axial ratio is anything other than 0 dB, the antenna is elliptically polarized, but it is common to refer to real-world antennas as circularly polarized with an axial ratio slightly greater than 0 dB.

When an antenna polarization is orthogonal to that of an incident wave, it will theoretically not receive any power from the incident wave. Examples of this are a vertically polarized wave incident on a horizontally polarized antenna or the Direct TV example of RCP and LCP waves being simultaneously incident on the antenna and only one or the other is actually sent to the receiver. Of course, real-world antennas and waves are not perfect, so it is valuable to be able to determine how much energy from an incident wave is coupled to the antenna. In the cross-polarization case, this is characterized by the cross-polarization discrimination (XPD). In the co-polarization scenario, if the polarizations do not match exactly, not all of the incident energy is coupled to the antenna. This is characterized by the polarization loss factor, F. Both of these parameters can be determined from the orientation and axial ratio of the wave and the antenna.

3.5.1 Cross-Polarization Discrimination

In some systems, orthogonal polarizations are used to provide two channels in the same frequency band. This may happen for example on a satellite communication system where both left and right circular polarizations are transmitted to the ground at the same frequency, and each earth station must receive the appropriate polarization. Any cross-polarization leakage is considered co-channel interference. Cross-polarization discrimination or isolation can be computed for either linear or circular polarizations. For fixed terrestrial links, it is possible to employ polarization diversity to increase frequency reuse, since the orientation of the polarization vectors can be carefully controlled. For satellite communication systems, polarization diversity requires the use of circular polarization since absolute control of the polarization vector orientation is difficult, due to the geometry and the potential for Faraday rotation in the ionosphere.

An important distinction that must be made is that technically the cross-polarization discrimination is between the incident wave and the receive antenna, not between the transmit and receive antennas [17]. The distinction is due to the fact that there may be environmental effects on the transmitted wave prior to reception. In most practical applications, however, the parameters of the transmit antenna are used either directly or with some allowance for environmental effects. For the linear polarization case, the cross-polarization discrimination is a function of the angle difference between the incident wave polarization vector and the receive antenna polarization vector. This is referred to as the tilt angle and often designated as τ. The amount of cross-polarization discrimination is then given by

$$XPD = \sin^2(\tau)$$

Cross-polarization discrimination for linear polarization can be thought of as a special case of circular or elliptical polarization. For elliptical polarization, the cross-polarization discrimination is a function of the tilt angle τ and of the axial ratios of the incident wave and the receive antenna. For linear polarization, the axial ratio is ideally infinite since the minor axis of the ellipse is zero. Thus linear polarization and circular polarization can both be viewed as special cases of elliptical polarization.

3.5.2 Polarization Loss Factor

The polarization loss factor is the multiplier that dictates what portion of the incident power is coupled into the receive antenna. The polarization loss factor will often be less than one because of polarization mismatch. The received power is given by

$$P_r = FP_i$$

where P_i is the incident power and P_r is the power actually available at the antenna. The polarization loss factor is the complement of cross-polarization discrimination. Just as with XPD, the polarization loss factor for linear polarization is a degenerate case of the elliptical polarization case, where the polarization ellipse collapses into a line. For linear polarization,

$$F = \cos^2(\tau)$$

where τ is the angle between the wave polarization and the receive antenna polarization. F is a power ratio, so it is converted to dB using $10\log(F)$. It is apparent that when τ is zero, $F = 1$ and all of the incident power is received.

When τ is 90 degrees the polarization of the wave and the receive antenna are orthogonal and no power is transferred. From the expressions for XPD and F for linear polarization, it is clear that

$$XPD = 1 - F$$

as expected based on conservation of energy.

The general development of polarization loss factor is based on elliptical polarization. Kraus [19] provides a development of the polarization loss factor as a function of the axial ratios and tilt angle using the Poncaire' sphere, the results of which are summarized here. The coordinates of the polarization vector on the sphere are

$$\text{Longitude} = 2\tau$$

$$\text{Latitude} = 2\varepsilon$$

where τ is the tilt angle as defined earlier and

$$\varepsilon = \tan^{-1}(k/AR)$$

with

$k = 1$ for left-hand polarization
$k = -1$ for right-hand polarization

The great-circle angle to the polarization vector is given by 2γ, where

$$\gamma = \frac{1}{2}\cos^{-1}(\cos(2\varepsilon)\cos(2\tau))$$

and the equator-to-great-circle angle is δ as previously defined. The polarization loss factor is defined as

$$F = \cos^2(\gamma_r - \gamma_w)$$

where $2\gamma_r$ and $2\gamma_w$ are the great-circle angles to the receive antenna polarization vector and the wave polarization vector, respectively.

$$\gamma_r = \frac{1}{2}\cos^{-1}(\cos(2\varepsilon_r)\cos(2\tau_r))$$

$$\gamma_w = \frac{1}{2}\cos^{-1}(\cos(2\varepsilon_w)\cos(2\tau_w))$$

It is common practice to let either τ_r or τ_w be zero and the other term then accounts for the net tilt angle between the wave and the antenna. Making the appropriate substitutions, it is possible to derive an expression for polarization loss as a function of γ_r, γ_w, and τ:

$$F = \cos^2\left\{\tan^{-1}\left(\frac{k}{AR_w}\right) - \frac{1}{2}\cos^{-1}\left[\cos\left(2\tan^{-1}\left(\frac{k}{AR_r}\right)\right)\cos(2[\tau_w - \tau_r])\right]\right\} \quad (3.12)$$

An equivalent and perhaps easier-to-use formulation is given in Ref. 17:

$$F = \frac{(1+AR_w^2)(1+AR_r^2)+4AR_wAR_r+(1-AR_w^2)(1-AR_r^2)\cos(2[\tau_w - \tau_r])}{2(1+AR_w^2)(1+AR_r^2)} \quad (3.13)$$

In each case, the polarization loss factor applies to the received power. Some interesting observations can be made from the preceding expression. If both the incident wave and the antenna are circular $(AR = 1)$, then F defaults to unity as expected, regardless of the tilt angle. On the other hand, if one of the axial ratios is unity and the other is greater than one, there will be some polarization loss regardless of the tilt angle. If both the incident wave and the antenna have identical axial ratios (greater than one), then the polarization loss factor becomes a function of the relative tilt angle only and complete power transfer occurs only when $\tau = 0$ or 180 degrees. The minimum and maximum polarization losses occur at 0/180 and 90/270 degrees, respectively.

For the purpose of link budgeting, the worst-case polarization loss factor is employed. The worst case occurs when $\tau = 90$ degrees. Figure 3.12 (a) is a plot of the polarization loss factor in dB for a tilt angle of 90 degrees, as a function of incident axial ratio for several different antenna axial ratios. By selecting a curve with the receive antenna axial ratio and finding the wave (transmit) axial ratio along the x axis or vice versa, the polarization loss factor can be read from the y axis of the plot.

In some circumstances, a circularly polarized antenna might be used to receive a linearly polarized wave or vice versa. This may occur when an antenna serves multiple purposes, or if the orientation of the linear polarization is unknown, then use of a circular polarized receive antenna might be considered. For instance, the orientation of a linearly polarized antenna on an aircraft will vary with the aircraft's attitude. In this case, using a circularly polarized antenna at the other end of the communication link would eliminate the variability in signal strength at the expense of taking a constant loss in signal strength of about 3 dB. The polarization loss factor between linear and circular polarization is generally assumed to result in a 3-dB polarization loss factor, regardless of the orientation of the linearly polarized wave. However, if the circularly polarized antenna is not ideal (i.e., axial ratio greater than 1), then the actual polarization loss factor is a function of the tilt angle.

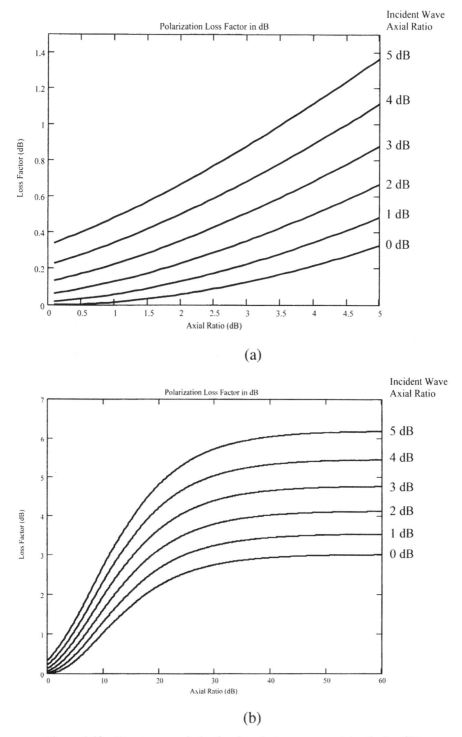

Figure 3.12   Worst-case polarization loss factor versus axial ratio for CP.

In this case, the polarization loss factor will be greater than or less than 3 dB depending upon the orientation and the axial ratio.

Figure 3.12 (b) shows the polarization loss factor for larger values of axial ratio along the x-axis. Since a linearly polarized antenna can be represented by an elliptical polarization with a large axial ratio, the far right-hand side of Figure 3.12 (b) shows how the worst-case polarization loss levels-off when one of the antennas tends toward linear polarization. From this plot, it can be seen that if the circularly polarized antenna (or wave) has an ideal axial ratio of 0 dB, then the worst-case polarization loss is 3 dB. Otherwise, if the circularly polarized antenna (or wave) is not ideal, then the actual circular-to-linear polarization loss factor exceeds 3 dB as shown. A more precise estimate of the polarization loss factor can be obtained by taking the limit of equation (3.12) as either AR_w or AR_w go to infinity, if desired.

Example 3.5. A desired satellite signal is transmitted with right "circular" polarization from an antenna having a 2-dB axial ratio. If the ground station receive antenna has an axial ratio of 3 dB, what are the minimum and maximum polarization loss that can be expected if environmental effects are negligible? If another signal is simultaneously transmitted from the same satellite antenna at the same frequency, but with left circular polarization, what is the worst-case cross-polarization isolation that will occur?

The maximum polarization loss factor can be read from Figure 3.12 and is

$$F_{max} = -0.35 \, dB$$

The minimum can be found by setting τ equal to 0 or 180 degrees and evaluating either of the expressions for axial ratio. The resulting minimum polarization loss factor is

$$F_{min} = -0.01 \, dB$$

which is negligible.

The cross-polarization isolation can be determined by using one minus the worst-case polarization loss factor:

$$XPD_{min} = 10\log(1 - 10^{F/10}) = -11.1 \, dB$$

Thus the interfering cross-polarized signal will be only 11 dB below the desired signal. □

3.6 ANTENNA POINTING LOSS

In applications where directional antennas are used, it is customary to allow for some antenna misalignment loss when computing the link budget. Antenna

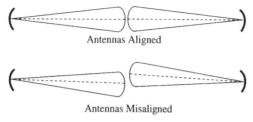

Figure 3.13 Antenna alignment and misalignment for a point-to-point link using directional antennas.

misalignment means that the received signal is not received at the peak of the antenna beam and/or the portion of transmit signal that is received did not come from the peak of the transmit antenna beam. This is shown pictorially in Figure 3.13. While the potential for misalignment is intuitive for mobile systems or systems tracking mobile satellites, it is also an issue for static links. Pointing loss can occur in several ways. The initial antenna alignment is seldom perfect; on a point-to-point link, each end must be aligned. In addition, effects such as wind, age, and thermal effects may all contribute to changing the antenna alignment over time. Variation in atmospheric refraction can produce time-varying pointing error, which can be challenging to identify.

For systems with active tracking on one end of the link, the tracking loss must take into account the track loop errors as well as any error introduced by the tracking process. A track loop must move the peak of the antenna beam about the actual track point to generate a tracking error signal for position correction. This process of scanning will result in a certain amount of signal loss (scan loss) that must be factored into the link budget. Most tracking loops require 0.5 to 1 dB of tracking signal variation for stable tracking.

3.7 SUMMARY

The antenna serves as the interface between system electronics and the propagation medium. The principle of reciprocity states that an antenna performs the same whether transmitting or receiving, which simplifies analysis and measurement. Antenna gain and directivity define the amplitude of the antenna pattern in the direction of maximum radiation (or reception). The gain of an antenna can be estimated using the concept of effective area for aperture antennas or effective height for linear antennas. Antenna gain, mainlobe width, and sidelobe levels are key concerns in antenna design or analysis.

The performance of an antenna varies with the distance from the antenna. The antenna's far-field, where the distance is greater than twice the square of the antennas largest linear dimension divided by the wavelength, is most often the region where the parameters are defined and specified. The reactive near-field is defined as the region where the distance is less than the wavelength divided by two pi. In the reactive near-field, the antenna pattern does not apply, and in fact the illumination function of the antenna will be present. Objects in the reactive near-field will couple with the antenna and alter the far-field pattern. The radiating near-field is the region between the reactive near-field and the far-field. In the radiating near-field, the antenna pattern is starting to take shape, but the amplitude in a particular direction may still vary with distance from the antenna. In this region, the radiated wave front is still spherical, versus nearly planar in the far-field.

There are a wide variety of antenna types and development continues. For most applications, antennas are required to transmit/receive either linear (vertical or horizontal) or elliptical (right-hand or left-hand) polarization. Any mismatch between the received wave and the receive antenna may lead to signal loss and reduction in rejection of the orthogonal polarization at the receive side of the link. The polarization loss factor and cross-polarization discrimination are often included in link budgets, particularly for satellite links.

REFERENCES

1. W. L. Stutzman and G. A. Thiele, *Antenna Theory and Design*, 2nd ed., Wiley, Hoboken, NJ, 1998 pp. 404–409.
2. A. V. Oppenheim and R. W. Schafer, *Digital Signal Processing*, Prentice-Hall, Englewood Cliffs, NJ, 1975, pp. 239–250.
3. F. J. Harris, On the use of windows for harmonic analysis with the discrete fourier transform, *Proceedings of the IEEE*, Vol. 66, No. 1, January 1978, pp. 51–83.
4. M. I. Skolnik, *Introduction to Radar Systems*, 3rd ed., McGraw-Hill, New York, 2001, p. 541.
5. W. L. Stutzman and G. A. Thiele, *Antenna Theory and Design*, 2nd ed., Wiley, Hoboken, NJ, 1998, p. 77.
6. J. D. Kraus and R. J. Marhefka, *Antennas for All Applications*, 3rd ed., McGraw-Hill, New York, 2002, pp. 30–36.
7. J. D. Kraus and R. J. Marhefka, *Antennas for All Applications*, 3rd ed., McGraw-Hill, New York, 2002, p. 31.
8. C. A. Balanis, *Antenna Theory, Analysis and Design*, 2nd ed., Wiley, New York, 1997, p. 164.
9. M. I. Skolnik, *Introduction to Radar Systems*, 3rd ed., McGraw-Hill, New York, 2001, p. 625.
10. *The ARRL Handbook for Radio Amateurs*, ARRL, Newington, CT, 1994, Table 41, Chapter 35, p. 38.
11. *The ARRL Handbook for Radio Amateurs*, ARRL, Newington, CT, 1994, Figure 24, p. 16-5.
12. C. A. Balanis, *Antenna Theory, Analysis and Design*, 2nd ed., Wiley, New York, 1997, pp. 32–34.

13. K. Siwiak, *Radiowave Propagation and Antennas for Personal Communications*, 2nd ed., Artech House, Norwood, MA, 1998, pp. 352–353.

14. W. L. Stutzman and G. A. Thiele, *Antenna Theory and Design*, 2nd ed., Wiley, Hoboken, NJ, 1998, pp. 187–195.

15. J. Liberti, Jr. and T. S. Rappaport, *Smart Antennas for Wireless Communications: IS-95 and Third Generation CDMA Applications*, Prentice-Hall, Upper Saddle River, NJ, 1999.

16. J. D. Kraus and R. J. Marhefka, *Antennas for All Applications*, 3rd ed., McGraw-Hill, New York, 2002, p. 45.

17. J. S. Hollis, T. J. Lyon, and L. Clayton, *Microwave Antenna Measurements*, 2nd ed., Scientific Atlanta, Atlanta, GA, 1970, Chapter 3.

18. J. D. Kraus and R. J. Marhefka, *Antennas for All Applications*, 3rd ed., McGraw-Hill, New York, 2002, p. 46.

19. J. D. Kraus and R. J. Marhefka, *Antennas for All Applications*, 3rd ed., McGraw-Hill, New York, 2002, pp. 47–52.

EXERCISES

1. What is the approximate gain of a circular dish antenna with 30-cm diameter at 40 GHz? Assume a total efficiency of 65% and give your answer in dB.

2. What is the approximate gain of a rectangular antenna dish with dimensions 100 cm by 20 cm at 10 GHz? Assume a total efficiency of 60%.

3. If a system can tolerate 0.5 dB of antenna mismatch loss, what are the minimum and maximum acceptable antenna driving point impedances? Assume that the system's characteristic impedance is 50 ohms.

4. Generate a plot of antenna mismatch loss versus VSWR.

5. What is the minimum and maximum possible polarization loss factor for a point-to-point communication system if each of the antennas used has an axial ratio of 2 dB. You may neglect environmental effects.

6. What are the effective height and effective area of a 0.1 wavelength dipole antenna if the antenna is resonant at 80 MHz?

7. If an aperture antenna has −17-dB sidelobes and the EIRP is 100 dBW, what is the worst-case power density that will be received at a distance of 2 m from the antenna if the receiver is outside of the transmit antenna's mainlobe?

CHAPTER 4

Communication Systems and the Link Budget

4.1 INTRODUCTION

This chapter treats the basics of communication systems analysis. It is valuable to design a communication system for the intended environment and to predict, with relative certainty, the system performance prior to fabrication and deployment. Radio-frequency (RF) planning for networks such as cellular telephone systems or wireless local area networks (LANs) is a key part of network deployment. Insufficient planning can result in overdesign and wasted resources or under design and poor system performance. Prior to planning a network, the parameters controlling the performance of each individual link must be understood. The essential parameters are the received signal strength, the noise that accompanies the received signal, and any additional channel impairments beyond attenuation, such as multipath or interference.

For link planning, a *link budget* is prepared that accounts for the transmitter effective isotropically radiated power (EIRP) and all of the losses in the link prior to the receiver [1]. Depending upon the application, the designer may also have to compute the noise floor at the receiver to determine the signal level required for signal detection. The link budget is computed in decibels (dB), so that all of the factors become terms to be added or subtracted. Very often the power levels are expressed in dBm rather than dBW due to the power levels involved. It is important that consistency of units be maintained throughout a link budget analysis.

The *link margin* is obtained by comparing the expected received signal strength to the receiver sensitivity or threshold (in satellite systems, the term sensitivity has a different meaning). The link margin is a measure of how much margin there is in the communications link between the operating point and the point where the link can no longer be closed. The link margin can be found using

$$\text{Link margin} = \text{EIRP} - L_{Path} + G_{Rx} - \text{TH}_{Rx}$$

Introduction to RF Propagation, by John S. Seybold
Copyright © 2005 by John Wiley & Sons, Inc.

where

EIRP is the effective isotropically radiated power in dBW or dBm

L_{Path} is the total path loss, including miscellaneous losses, reflections, and fade margins in dB

G_{Rx} is the receive gain in dB

TH_{Rx} is the receiver threshold or the minimum received signal level that will provide reliable operation (such as the desired bit error rate performance) in dBW or dBm

The available link margin depends upon many factors, including the type of modulation used, the transmitted power, the net antenna gain, any waveguide or cable loss between transmitter and antenna, radome loss, and most significantly the path loss. The modulation affects the link margin by changing the required E_b/N_0* or SNR. The antenna gains, transmission losses, and transmitted power all directly affect the link budget. The path loss is the most significant factor because of its magnitude relative to the other terms. The path loss includes geometric spreading loss or free-space loss (FSL) as well as environmental factors. The term path loss is sometimes used to refer to free-space loss and sometimes refers to the entire path loss experienced by a communication link. The intended meaning should be clear from the context.

4.2 PATH LOSS

Path loss is the link budget element of primary interest in this text as it relates to the topic of RF propagation. The path loss elements include free-space loss (if appropriate; otherwise a median loss is computed using other methods as discussed in later chapters), atmospheric losses due to gaseous and water vapor absorption, precipitation, fading loss due to multipath, and other miscellaneous effects based on frequency and the environment.

If the principal path is governed by free-space loss, it is calculated using the Friis free-space loss equation, which can be expressed as

$$L = G_T G_R \left(\frac{\lambda}{4\pi d} \right)^2 \qquad (4.1)$$

As shown here, L is actually a gain (albeit less than unity) rather than a loss. This will be addressed shortly. Note that the Friis equation includes transmit and receive antenna gain. In certain applications, the exponent in the free-space loss equation is a value other than 2. This reflects the geometry of the environment and is covered in later chapters. If a so-called modified power

---

* E_b/N_0 is the energy per bit divided by the noise power spectral density (see Section 4.5.5).

law is used, then, strictly speaking, it is not free-space loss, but rather represents a mean or median path loss in a lossy environment such as near-earth propagation. The examples in this chapter are restricted to square-law free-space loss.

The antenna gain accounts for the antenna directivity and efficiency, while the inverse distance squared term accounts for the spherical wave-front (geometric) spreading. The dependence on wavelength is an artifact of using the receive antenna gain in the equation rather than the antenna effective area.

The Friis free-space loss equation can be expressed in dB as

$$L_{dB} = -G_{TdB} - G_{RdB} - 20 \log(\lambda) + 20 \log(d) + 22 \qquad (4.2)$$

where a negative sign has been included so that the value of L_{dB} is, in fact, a loss. One way to look at this is that a loss is simply the negative of a gain when expressed in dB. While the subtleties of this are not essential, keeping the signs straight when performing a link budget certainly is.

In many applications, the antenna gains are excluded from the path loss expression, in which case the free-space loss is

$$L_{FSL\ dB} = -20 \log\left(\frac{\lambda}{4\pi d}\right) \qquad (4.3)$$

Example 4.1. Consider a 100-m link that operates in free space at 10 GHz. Assume that the transmit power is 0.1 W, and both transmit and receive antennas have 5-dB gain. If the receiver threshold is −85 dBm, what is the available link margin?

The first step is to compute the EIRP of the transmitter. Since the receive threshold is given in dBm, it is helpful to have EIRP in dBm as well.

$$P_T = 100 \text{ mW} = 20 \text{ dBm}$$

Then, since

$$\text{EIRP} = P_T G_T \qquad \text{mW}$$

or, in dB

$$\text{EIRP}_{dB} = P_{TdB} + G_{TdB} = 25 \text{ dBm}$$

and the path loss is

$$PL = -20 \log\left(\frac{\lambda}{4\pi d}\right) = 92.4 \text{ dB}$$

The resulting link margin is

$$M_L = 25 - 92.4 + 55 - (-85) = 22.6 \text{ dB} \quad \square$$

There are additional sources of signal loss within the transmitter and receiver. These losses must also be accounted for (as was done in Example 4.1) in order to accurately predict the received signal level and the resulting signal-to-noise ratio. Sometimes the magnitude of these effects will be specified by the equipment designer or manufacturer. When they are not known, an experienced analyst can estimate them reasonably well.

Some of the additional loss factors include band-limiting and/or modulation loss. Transmit filters control spectral spreading but add insertion loss, while some types of digital modulation have lower efficiency due to transmitted carrier components (i.e., the signal power is somewhat less that the bit energy times the bit rate, $E_b R_b$).

If there is a radome covering the antenna, it may reduce the effective antenna gain and increase antenna noise. This problem is exacerbated when the radome is dirty or wet. In addition, any waveguide or cable used to connect to the antenna is a source of loss and of additional noise. It is important that waveguides be intact and fitted properly and that coaxial cables be appropriate for the frequency and be in good condition. Kinked or deteriorated coaxial cable can produce significant loss at either the transmitter or the receiver.

Point-to-point links often use directional antennas to increase the energy directed at the receiver. This can either increase range or reduce the required transmit power to close the link. If the directional antennas are not perfectly aligned, however, the effective antenna gain is reduced. Depending upon the application, a dB or two may be allocated to pointing loss in the link budget. If there is a directional antenna on each end of the link, there will be some pointing loss allocated to transmit and some to receive. Another loss contributor is polarization loss that depends upon the relative orientation of the transmit and receive polarization vectors and the axial ratio of each antenna. A mismatch in the transmit and receive polarization vector alignment reduces energy transfer and becomes a source of loss. This is of minor importance on fixed, point-to-point links, but is a consideration for mobile applications and satellite systems. Finally, atmospheric loss due to such factors as rain, water vapor absorption, and oxygen absorption can be significant for long links at high frequencies. A detailed discussion concerning the prediction of rain attenuation is presented in Chapter 10, and a discussion of atmospheric losses is presented in Chapter 6.

4.3   NOISE

The key receiver characteristic that determines link performance is the receive threshold or sensitivity. The threshold is the minimum signal strength required for acceptable operation. If the threshold is not known or specified, then a required signal-to-noise ratio or bit energy to noise power spectral density ratio (E_b/N_0) for desired operation or probability of detection must be provided. For most terrestrial systems, the noise is set by the receiver front-end characteristics and the bandwidth. For satellite ground station receivers, the antenna temperature may be very low, which will reduce the noise floor. Since the antenna temperature for a skyward looking antenna may be less than 290 K, losses in the path may reduce the signal level and increase the noise temperature. This is discussed in Chapter 11. For the purpose of this chapter, all losses and antennas are at 290 K, so the effect of losses is simply to reduce the signal level while the noise stays at the same level.

Receiver noise is caused by the thermal agitation of electrons, called *Brownian motion*, in the front end of the receiver [2]. This noise is modeled as having a Gaussian amplitude distribution. In addition, it is modeled as having a flat (constant) power spectral density of $N_0/2$, meaning that the power density is the same over all frequencies of interest. Thus thermal noise usually characterized as "white". Finally, the thermal noise is additive in nature (i.e., independent of the actual signal involved). So thermal noise is modeled as additive white Gaussian noise (AWGN).

The theoretical thermal noise power available to a matched load is

$$N = kT_0B \qquad (4.4)$$

where

$$k = 1.38 \times 10^{-23} \text{ J/K} \qquad \text{(Boltzman's constant)}$$

$$T_0 = 290 \text{ K} \qquad \text{(standard noise temperature)}$$

$$B = \text{noise-equivalent bandwidth (Hz)}$$

This represents the amount of noise power present at the input to a perfect receiver that is driven by a matched load (such as an antenna).

The noise power available at the output of a linear filter is given by [3]

$$N = \frac{1}{2\pi} \int_{-\infty}^{\infty} G_N(f)|H(f)|^2 \, df \qquad (4.5)$$

Since the power spectral density of the noise is constant for AWGN, it can be brought out of the integral. Since

$$G_N(f) = N_0/2 = kT_0$$

(4.5) can be rewritten as

$$N = \frac{N_0}{4\pi} \int_{-\infty}^{\infty} |H(f)|^2 \, df \qquad (4.6)$$

Using the fact that the integrand in the above equation will be even for real filters, the equation for the total noise power becomes

$$N = N_0 B |H(f_0)|^2$$

where B is the noise equivalent bandwidth, defined by [4]

$$B = \frac{\int_0^\infty |H(f)|^2 \, df}{|H(f_0)|^2} \tag{4.7}$$

and N_0 is twice the value of the (two-sided) noise power spectral density. Note that the lower limit of integration was changed to accommodate the factor of two that was included with the noise power spectral density. Figure 4.1 shows the spectra of the noise power density, a representative band-pass character-istic and the corresponding ideal bandpass filter. The noise equivalent band-width is the width of the ideal bandpass filter as shown. Often the filter is treated as unity gain, so that $|H(f_0)|^2 = 1$. The actual gain or loss of the filter is then included elsewhere in the computation.

The noise-equivalent bandwidth is the bandwidth of an ideal bandpass filter that would pass the same amount of white noise power as the system being analyzed. In most modern digital communication systems, it is closely approxi-mated by

$$B_N = \frac{1}{T_S} \quad \text{Hz} \tag{4.8}$$

where a root-raised cosine or similar filter used to shape the symbol and T_S is the symbol duration. The 3-dB bandwidth is sometimes a good approximation to the noise equivalent bandwidth and is often used as such, particularly when detailed information on the system response is not available. Modern test instruments provide the capability to rapidly measure the noise-equivalent bandwidth, which greatly simplifies the task.

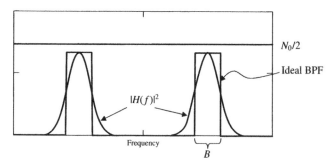

Figure 4.1 Noise power spectral density along with actual and ideal filter responses.

Real-world components are sources of additional noise so that the overall noise level is always greater than the theoretical value. For example, a receiver adds noise of its own by the very presence of resistive elements in its front end. Receiver noise can be characterized by either the effective noise temperature or the noise figure (they are related). The actual noise power is then

$$N = kT_0B + kT_eB \qquad \text{watts}$$
$$N = kT_0BF \qquad \text{watts}$$

(4.9)

where $F = (1 + T_e/T_0)$ is called the noise factor and T_e is the equivalent noise temperature of the receiver. This term, F, is generally referred to as the noise figure, although, strictly speaking, the noise figure is the noise factor expressed in dB. This misnomer is perpetuated in this text in conformance with common usage. The reader should be sure to understand the distinction, however.

The computation of the total noise power can also be performed in dB:

$$N_{dBW} = -204 \text{ dB- W}/\text{Hz} + 10 \log(B)\text{dB-Hz} + F_{dB} \qquad (4.10a)$$

or

$$N_{dBm} = -174 \text{ dBm}/\text{Hz} + 10 \log(B)\text{dB-Hz} + F_{dB} \qquad (4.10b)$$

where F_{dB} is the noise figure. The noise figure can also be defined as the ratio of the input signal-to-noise ratio to the output signal-to-noise ratio. The two definitions are equivalent, which is readily shown. Let

$$F = \frac{\text{SNR}_{in}}{\text{SNR}_{out}} \qquad (4.11)$$

The signal-to-noise ratios are

$$\text{SNR}_{in} = \frac{S}{kT_0B}$$
$$\text{SNR}_{out} = \frac{S}{k(T_0 + T_e)B}$$

Substituting these expressions into (4.11) yields

$$F = 1 + \frac{T_e}{T_0} \qquad (4.12)$$

Example 4.2. Given a receiver for a 10-megasymbol-per-second signal, determine the available noise power present with the signal and find the noise figure of the receiver if the equivalent noise temperature of the receiver is 870 K.

A 10 megasymbols per second data rate implies a noise-equivalent bandwidth of approximately 10 MHz. The noise figure is determined using (4.12) and is 4 (6 dB). The noise power present with the received signal will be

$$N = kT_0BF = -128 \text{ dBW} \quad \square$$

When there are several cascaded elements to be considered, the noise figure of the composite system can be readily determined. First, consider the case of two cascaded amplifiers as shown in Figure 4.2. Assume the system is driven by a source whose impedance matches the first amplifier stage and whose noise level is

$$N_{in} = kT_0B$$

The accompanying signal is S_{in}. The output of the first stage will be

$$S_1 = G_1S_{in} + G_1N_{in} + G_1kT_{e1}B$$

or

$$S_1 = G_1S_{in} + G_1kT_0BF_1$$

where T_{e1} is the effective temperature of the first stage, referenced to the input. The signal at the input to the second amplifier is the output of the first amplifier, S_1. To determine the output of the second amplifier, we add the apparent noise of the second amplifier, which is characterized by $T_0 + T_{e2}$, and multiply by the gain. Thus the output of the second amplifier is

$$S_2 = G_2S_1 + G_2G_1kT_0BF_1 + G_2k(T_0 + T_{e2})B$$

or

$$S_2 = G_1G_2S_{in} + G_1G_2kT_0BF_1 + G_2kT_0BF_2$$

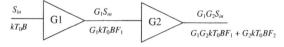

Figure 4.2 Noise figure of two cascaded amplifiers.

To find the noise figure of the combined system, the definition of the noise figure is applied:

$$F = \frac{(S/N)_{in}}{(S/N)_{out}}$$

so

$$F_{total} = \frac{S_{in}/kT_0B}{G_1G_2S_{in}/[G_1G_2kT_0B(F_1 + F_2/G_1)]}$$

or

$$F_{total} = F_1 + \frac{F_2}{G_1} \tag{4.13}$$

Note that if

$$G_1 \gg F_2$$

then

$$F_{total} \simeq F_1$$

By applying mathematical induction, it can be shown that for any number of cascaded amplifiers, the composite noise figure is given by

$$F_{total} = F_1 + \frac{F_2}{G_1} + \frac{F_3}{G_1G_2} + \frac{F_4}{G_1G_2G_3} + \dots \tag{4.14}$$

Based on the preceding analysis, it can be seen that the first stage of a well-designed system sets the noise floor and thus must have the best noise performance of all of the stages and relatively high gain. It is also apparent that any losses prior to the first gain stage (usually a low-noise amplifier or LNA) increase the noise figure. This can be understood by remembering that the noise floor is set by the receiver characteristics (or the LNA) and thus the noise is not reduced by the loss, but the signal is reduced; consequently, the signal-to-noise ratio at the output is reduced, which decreases the noise figure.

Any losses present between the antenna output and the receiver input increase the overall noise figure of the system. For terrestrial systems, where the input temperature is assumed to be the standard noise temperature of 290 K, the new noise figure is computed by simply adding the losses to the receiver noise figure. This is seen in the following example.

Example 4.3. Consider the receiver system shown below.

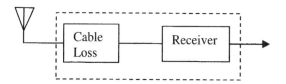

Given that the cable loss is 7 dB and the noise equivalent temperature of the receiver is 630 K, what is the noise figure of the receiver system?

The noise figure of the receiver alone is

$$F = 1 + \frac{T_e}{T_0} = 5 \, \text{dB}$$

So combining the receiver noise figure with the loss that precedes the receiver, the noise figure of the entire receiver system is computed to be 12 dB.

Looking at the same problem from the signal-to-noise ratio perspective, let the signal at the output of the antenna be denoted by S_1 and let the noise density at the output of the antenna be

$$N_1 = kT_0$$

The signal-to-noise ratio at the input to the cable is S_1/N_1. The 7-dB cable loss means that only 20% of the input power arrives at the output. Denoting the signal at the output of the cable as S_2,

$$S_2 = 0.2 \, S_1$$

The noise density at the output of the cable is based on the standard noise temperature (since the cable is also a noise source), so

$$N_2 = kT_0 = N_1$$

Thus the noise figure of the cable can be found:

$$F_{cable} = (S_1/N_1)/(S_2/N_2) = 5$$

or 7 dB as expected. Treating the receiver as a unity gain amplifier, one may apply the cascade formula for noise figure to determine the overall noise figure, which again is shown to be 12 dB. $\square$

It is important that only absorptive (resistive) losses be included in the system noise figure computation. Gain reductions such as pointing error are not treated as noisy attenuators since the noise is referenced to the input of the receiving system.

There are many other sources of noise and loss that may be relevant to a given application. Some examples are:

- *Local Oscillator (LO) Phase Noise.* LOs are references for modulation and demodulation of an information signal. Phase jitter within the LO causes unintended spreading of the signal energy at both the transmission and receiving ends of a link. In most cases, spread components cannot be recovered and may accounted for as a loss. In-band components are seen as an increase in noise level.

- *AM-to-AM and AM-to-PM Conversion.* System nonlinearities in the transmitter or receiver may cause spurious amplitude variations and/or phase variations in the signal. Depending on the nature of the nonlinearities, some of the signal may be spread out-of-band, resulting in an overall signal loss.

- *Intermodulation Products.* Some systems may use several carriers that are amplified by a single wideband transmitter amplifier. Nonlinearities in the amplifier result in the various carriers being multiplied, thereby creating new signals at the sum and difference frequencies and at all combinations of the carrier frequencies. Here, too, the effects may be overall signal loss and/or added in-band noise or loss.

- *Synchronization Loss.* Imperfect synchronization results in degradation in detector performance, and therefore a loss, and may also introduce additional noise.

These noise sources are, in general, neither white nor Gaussian and may be multiplicative rather than additive. Depending on the nature of these effects, they may be accounted for as a system loss only—that is, an increase in the minimum required E_b/N_0, RSL or received SNR.

The approach discussed in this chapter includes a standard noise temperature at the input. In satellite applications, the receiver may be sufficiently sensitive that it sees the sky noise from the antenna rather than a 290 K noise floor. In such applications, it is customary to use a total system temperature rather than an effective temperature or a noise figure. This is covered in Chapter 11, "Satellite communications."

4.4 INTERFERENCE

Bandwidth limitation in the transmitter, receiver, and/or the channel itself may cause pulse overlapping in time. Typically, most digital communication systems have very stringent bandwidth allocations, and thus the digital pulses must be carefully shaped at the transmitter to avoid channel spillover. As the filter bandwidth becomes narrower and the cutoff sharper, the pulses tend to spread in time, causing them to overlap. This is called intersymbol interference (ISI) and can seriously degrade performance.

If multiple carriers are transmitted from the same transmitter, the nonlinearities can produce intermodulation products, which may be viewed as additional noise. Cripps [5] provides a good treatment of this subject. A similar effect occurs if there is a strong adjacent channel signal overloading the receiver and producing nonlinear operation [6]. In either case, the analysis (or testing) is usually performed using two closely spaced tones of equal amplitude. The nonlinearity produces spurs at various multiples, sums, and differences of the original two frequencies.

Another source of interference is external interference from other transmitters. If the interference is at the same frequency as the signal of interest, it is called *co-channel interference*. Co-channel interference may be caused by harmonics from a different type of system, unintentional radiators, or signals from a similar system that are some distance away (frequency sharing). In each case, the interference is received within the operating bandwidth of the receiver as shown in Figure 4.3.

Interference that is near the frequency of the signal is called *adjacent channel interference* and can be a problem depending upon the spectral properties of the receiver filter. If the filter skirts are not sufficiently attenuated, the interference can cause undesired operation. This is shown graphically in Figure 4.3. The normalized receiver filter response and its 3-dB bandwidth are indicated along with representative spectra for co-channel and adjacent channel interference. Since the co-channel interference enters the receiver at or near the center of its bandwidth, the receiver filter does not attenuate it. On the other hand, adjacent channel interference enters the receiver at a nearby frequency and will be attenuated by the skirts of the receiver filter. Thus a sharper roll-off on the receiver filter will reduce vulnerability to adja-

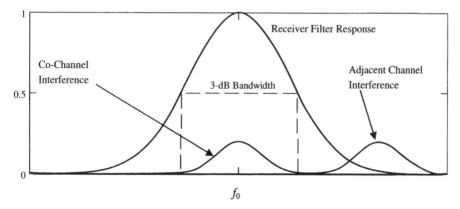

Figure 4.3   Receiver filter response with co- and adjacent channel interference.

cent channel interference, at the risk of possibly increasing the intersymbol interference.

It is possible that interference can be of such a nature that it appears to be externally generated noise. This is often how interference analysis is performed. Most of the time, however, interference is not noise-like, which makes precise performance analysis difficult. An exception is interference from specific systems, which can be carefully characterized on the bench to provide some understanding of the signal interactions for different relative powers, waveform timings, and frequency separations.

An interference margin term is included in the link budget. This term provides margin for noise floor degradation due to external interference. A 1-dB interference margin means that the noise floor (noise plus interference) will be as much as 1 dB higher than in a clear environment. The interference analysis can then be performed on a link with the condition that the interference must not degrade the noise floor by any more than 1 dB unless the interference margin is increased in the link budget.

Example 4.4. Given a system whose link budget includes a 1-dB interference margin, what is the maximum interference power at the input to the receiver that can be tolerated?

If the noise floor (kT_0BF) is N dBm, then the total noise plus interference can be as large as $N + 1$ dBm. Of course the powers must be added in milliwatts or watts, not dBm or dB. Let

$$n = 10^{N/10} \text{ dBm}$$

and

$$i = 10^{I/10} \text{ dBm}$$

where

n is the noise power in milliwatts
N is the noise power in dBm
i is the interference power in milliwatts
I is the interference in dBm

Then,

$$N + 1 = 10 \log(n + i)$$

Solving for i, yields

$$i = 10^{(N+1)/10} - n$$

$$i = n(10^{0.1} - 1) = 0.259n$$

Converting back to dBm, the equation for the interference power is,

$$I = N - 10 \log(2.59) = N - 5.9 \text{ dBm}$$

Thus a 1-dB interference margin requires that the total of all interference power remains 5.9 dB or more below the noise floor. This is a useful rule of thumb. □

Intersymbol interference is another channel effect that can degrade the performance of a digital communication system [7]. In the context of digital communications, the *channel* refers to everything that is between the transmitting and receiving modem, including RF mixers and other components, antennas and cabling, and the actual propagation medium. Thus the *channel* is a superset of the propagation channel. Inter-symbol interference may be generated in the transmitter, the receiver, the propagation channel, or any combination thereof. The propagation mechanisms that produce ISI are discussed in later chapters. The transmitter and receiver mechanisms that contribute to ISI are related to spectral shaping. Improper transmit or receive spectral shaping can produce symbols that overlap and interfere in time. The allocated margin for ISI is usually included in the implementation loss figure and addressed as part of modem design. For this reason, the ISI is not usually addressed by the link designer as a source of noise.

It is possible for a nearby strong signal to interfere with receiver operation even if it is well-filtered by the receiver and it does not overload (saturate) the front end LNA. This can occur if the interference is outside of the receiver's instantaneous (detection) bandwidth, but within the tuning bandwidth of the receiver. In this circumstance, the interfering signal may capture or drive the front-end's automatic gain control (AGC). When this occurs, the receiver's front-end gain is reduced to properly scale the interfering signal, which can attenuate the desired signal. This is called de-sensing and must be considered in configurations where other transmitters will be situated near the receiver.

4.5 DETAILED LINK BUDGET

The link budget is the compilation of all gains and losses in the communication link [1]. By summing the transmit power in dBm (or dBW) with all of the relevant link gains and losses in dB and then subtracting the required received signal level (RSL, expressed in the same units as the transmit power), the link margin in dB is obtained. The link margin provides a measure of robustness for a link. Links with relatively small margins are not likely to be very robust unless all of the gains and losses are very well understood and modeled. Some of the losses in a link budget are based on availability requirements, and therefore the actual losses are not always present. Availability requirements can drive some of the fade margins that are included in the link budget. In such circumstances a designer must account for losses that may be present inter-mittently such as the attenuation due to rain or other precipitation in order to ensure that the link is in fact available the required percentage of time. It is important that the required fade margin be preserved if the true desired availability is going to be maintained. This is discussed further in Chapter 10 on rain attenuation.

Figure 4.4 shows a typical link budget for a terrestrial millimeter-wave link. It is customary to include relevant parameters in the link budget so that a reader can reproduce the computations if desired. Thus, the frequency and antenna polarization are included in the first few lines. Link budget formats tend to be a religion, with little agreement on what is ideal or even necessary. For that reason, the following example should be used as a guideline and not necessarily as a template. The following sections closely examine each section of the link budget in Figure 4.4, providing the details of the computations.

4.5.1 EIRP

After the first introductory lines of the link budget in 4.4, the next several lines are dedicated to computing the transmitted EIRP. The EIRP is the transmit power plus the antenna gain, minus any waveguide and/or radome losses. Thus

$$\text{EIRP}_{dB} = P_{TxdBm} + G_{TxdB} - L_{WGdB} - L_R \text{dBm}$$

An allocation for loss due to impedance mismatch between the antenna and the transmitter or HPA may also be included as a line item in this section, or it may be included within the transmitter or antenna parameters. See Section 3.2.5 for details on impedance mismatch loss. The radome loss term may vary, increasing when the radome is wet. The link budget must use the expected worst-case value since the state of the radome is not known or controlled.

For the example link budget in Figure 4.4 the EIRP is calculated as

$$\text{EIRP}_{dB} = 10 \text{ dBm} + 32 \text{ dB} - 2 \text{ dB} - 1.5 \text{ dB} = 38.5 \text{ dBm}$$

Of course the units of the EIRP will be the same as the units of the transmit-ter power (dBW or dBm).

4.5.2 Path Loss

The path loss section includes the free space loss and all environmental effects including multipath. Some of these effects are probabilistic while others are deterministic. The free-space loss (or other computation of the geometric loss) is a deterministic function of range. Atmospheric loss is a function of range, the elevation angle of the link, and the local environment. A standard atmosphere may be used to estimate the loss for all cases of interest with acceptable error except at frequencies where the loss is considerable (e.g., 22

Frequency	38.6 GHz	
Wavelength	0.0078 m	
Polarization	Vertical	
Link Distance	2 km	
Tx Power	10.0 dBm	
Tx Loss	−1.5 dB	
Tx Antenna Gain	32.0 dB	
Radome Loss	−2.0 dB	
EIRP	**38.5 dBm**	
Path Loss (FSL)	−130.2 dB	
Tx Pointing Error	−1.0 dB	
Rain Loss (0.999)	−15.0 dB	
Multipath	−2.0 dB	
Atmospheric Loss	−0.2 dB	Gamma = 0.12 dB/km
Total Path Losses	**−148.4 dB**	
Radome Loss	−2.0 dB	
Rx Antenna Gain	32.0 dB	
Polarization Loss	−0.2 dB	
Rx Loss	−2.0 dB	
Rx Pointing Error	−1.0 dB	
Total Rx Gain	**26.8 dB**	
RSL	**−83.1 dBm**	
Interference margin	−1.0 dB	
Rx Noise Figure	7.0 dB	
Noise Bandwidth	25.0 MHz	
Total Noise Power	**−93.0 dBm**	$10\log(kT_0BF)$
Signal-to-Noise Ratio	8.9 dB	
Threshold	−88.0 dBm	
Net margin	**3.9 dBm**	

Figure 4.4 Sample link budget.

and 60 GHz). See Chapter 6 for detailed information on modeling atmospheric losses.

The probabilistic losses must have a fade margin in the link budget and a corresponding probability that the fade margin is exceeded. The complement of that probability is the link availability due to that particular phenomenon. For instance, in this example, the probability of a rain fade exceeding 15 dB is 0.1% (corresponding to the link availability of 0.999). If the rain fade does exceed that value, it is assumed that the link will be temporarily unavailable. The chosen probability must be within acceptable limits for the application being considered. The allowable path loss is given by the sum (in dB) of the free-space loss, the budgeted fade margin, and any miscellaneous losses.

$$PL_{dB} = FSL_{dB} + FM_{dB} + L_{miscdB}$$

For this example, the path loss is given by

$$PL_{dB} = -20\log(1/(4\pi d)) + 30 + 0.5 = 135\,\text{dB}$$

4.5.3 Receiver Gain

The receiver gain is equal to the antenna gain less any radome loss, cable or waveguide loss (receiver loss), polarization loss, and pointing loss.

$$G_{RdB} = G_{RxdB} - L_{RadomedB} - L_{WGdB} - L_{Pol} - L_{Pt}$$

An allocation for loss due to impedance mismatch between the antenna and the receiver may also be included as a line item in this section (see Section 3.2.5 for details).

The polarization loss is due to the axial ratios of the transmit and receive antennas and their relative orientation. The polarization loss, discussed in Section 3.5.2, is estimated using the worst-case conditions and then treated as a constant since it is not usually large and since it will remain fixed for a given geometry.

For this link budget, the net receive gain is given by

$$G_{RdB} = 32 - 2 - 2 - 0.2 - 1 = 26.8\,\text{dB}$$

The receive signal level is the EIRP minus the path loss, plus the receiver gain:

$$RSL = EIRP - PL + G_{RdB}$$

The received signal level is

$$RSL = 27\,\text{dBm} - 135\,\text{dB} + 17\,\text{dB} = -91\,\text{dBm}$$

4.5.4   Link Margin

The result of a link calculation may be the expected received signal level (RSL), the signal-to-noise ratio (SNR), the carrier-to-noise ratio (CNR), or the bit energy to noise PSD ratio (E_b/N_0). If the RSL is determined, it can be compared to the *minimum detectable signal* (MDS), or the system threshold. For the link in Figure 4.4, the system threshold is subtracted from the RSL plus the interference margin to determine the link margin.

$$(\text{RSL} + M_{INT}) - \text{TH}_{Rx} = 3.9 \text{ dB}$$

An experienced designer can often compute a link budget that is within a fraction of a dB of measured results.

The transceiver designer's objective is to close the link for the specified distance and required availability as efficiently as possible, whereas the link designer's task is to use an existing transceiver to satisfy a specific communication requirement. The link designer may then trade off availability for link distance.

4.5.5   Signal-to-Noise Ratio

If the signal-to-noise ratio is required for the link budget, it can be computed as follows. The receiver noise is given by

$$N_{dBm} = 10 \log(kT_0 BF)$$

$$N_{dBm} = -174 \text{ dBm} + 10 \log(B) + F_{dB}$$

For this example, the noise power is

$$N_{dBm} = -174 + 60 + 7 = -93 \text{ dBm}$$

The signal-to-noise ratio can then be computed using the RSL and the noise floor.

$$\text{SNR} = \text{RSL}_{dBm} - N_{dBm} = 16 \text{ dB}$$

If the received SNR had been specified rather than the RSL, the link margin would be computed by subtracting the required receive SNR from the predicted SNR.

For digital communication systems, the E_b/N_0 of the received signal is the determining factor on signal detection and the bit error rate (BER). The required E_b/N_0 for a given BER is provided by the modem designer and must be met or exceeded by the receiver designer if satisfactory performance is to be achieved. The bit-energy-to-noise-density ratio can be determined from the signal-to-noise ratio if the data rate and symbol rate are known. The noise spectral density is the total noise power divided by the noise equivalent bandwidth. If the noise equivalent bandwidth is not known, but the symbol rate is, then the noise-equivalent bandwidth can be approximated by the symbol rate, R_s, in symbols per second.

$$B \approx R_s$$

Noise density can then be expressed as

$$N_0 = N/B$$

The energy per bit can be determined by dividing the signal power by the bit rate*:

$$E_b = S/R_b$$

where R_b is the bit rate in bits per second. Thus,

$$E_b/N_0 = \frac{S}{N} \frac{B}{R_b}$$

The appropriate value to use for the bit rate is sometimes a source of confusion. Due to the use of digital coding, the bit rate in the channel may be higher than the data bit rate at the baseband output of the modem. The bit rate used in the above calculation is the actual data bit rate, not the channel bit rate. This is because the required E_b/N_0 for the modem is specified based on the output data rate, which can be achieved using various channel rates by means of different modulation and coding schemes. Therefore the link designer need not even know what modulation or coding the modem is using.

4.6 SUMMARY

In order to characterize the expected performance of a communication system for a given link, a link budget can be prepared, which is a complete accounting of the system performance and the expected channel impairments. By computing the expected receive signal level and comparing it to the system threshold, a link margin is found, which is a measure of the robustness of the link.

The path loss section of a link budget represents the largest source of loss in the link. As used herein, path loss refers to the free-space loss and any other propagation impairments that affect the received signal level. In addition to the free-space loss, the path loss section of the link budget typically includes atmospheric losses, transmitter antenna pointing loss, and multipath. Some of

* The bit rate is equal to the symbol rate multiplied by the number of bits per symbol.

these losses are deterministic (based on range), whereas others such as rain are probabilistic. For the probabilistic losses, a fade margin is included in the link budget, which has an associated probability of not being exceeded. This results in an availability value for that impairment being attached to the link budget.

The receiver noise level is a function of the receiver noise figure, miscellaneous losses in the receiver, and sometimes losses within the atmosphere. The thermal noise in the receiver is computed using Boltzmann's constant, the standard noise temperature, the system noise figure, and the noise equivalent bandwidth. Certain types of interference may also be treated as noise sources. Most link budgets include a term to allow for some signal-to-noise degradation due to interference.

REFERENCES

1. B. Sklar, *Digital Communications Fundamentals and Applications*, 2nd ed., Prentice-Hall, Upper Saddle River, NJ, 2001, pp. 286–290.
2. F. G. Stremler, *Introduction to Communication Systems*, 3rd ed., Addison-Wesley, Reading, MA, 1992, pp. 188–208.
3. F. G. Stremler, *Introduction to Communication Systems*, 3rd ed., Addison-Wesley, Reading, MA, 1992, p. 197.
4. J. G. Proakis and M. Salehi, *Communication Systems Engineering*, 2nd ed., Prentice-Hall, Upper Saddle River, NJ, 2002, p. 190.
5. S. C. Cripps, *RF Power Amplifiers for Wireless Communications*, Artech House, Norwood, MA, 1999, Chapter 7.
6. R. E. Zeimer and R. L. Peterson, *Introduction to Digital Communication*, 2nd ed., Prentice-Hall, Upper Saddle River, NJ, 2001, pp. 79–82.
7. R. E. Zeimer and R. L. Peterson, *Introduction to Digital Communication*, 2nd ed., Prentice-Hall, Upper Saddle River, NJ, 2001, pp. 155–168.

EXERCISES

1. Consider a receiver with a 4-dB noise figure and a 2-MHz bandwidth. If a carrier-to-noise ratio of 15 dB is required for acceptable BER (bit error rate) performance, what carrier (signal) strength is required at the receiver? At what point would this signal be measured or referenced?

2. In the preceding problem, if the transmitter at the other end of the communications link is 2 km away, both the transmitter and receiver use a 4-dB antenna, and the frequency is 800 MHz, what is the minimum required transmit power?

3. Given a receiver with a 100-kHz bandwidth and an effective noise temperature of 600 K, what is the noise power level at the input to the receiver? Give the units with your answer.

4. Given a symmetrical line-of-sight communication link with a minimum detectable signal of −90 dBw, a transmit power of +10 dBw, antenna gain of 28 dB, and frequency of 10 GHz, mounted on a 100-ft tower, what is the maximum communication distance (neglecting all sources of interference or fading)?

5. Consider a point-to-point communication link operating at 38 GHz, using 35-dB dish antennas at each end. The radome loss is 2 dB at each end, and a 10-dB fade margin is required to allow for rain fades. If the transmit power is −10 dBm, the receiver noise figure is 7 dB, the bandwidth is 50 MHz, and the link distance is 1.2 km, what is the signal-to-noise ratio at the receiver in dB? Assume that free-space loss applies.

6. Consider a communication link operating at 28 GHz, using 30-dB dish antennas at each end. A 10-dB fade margin is required to allow for rain fades. If the transmit power is 0 dBm, the receiver noise figure is 7 dB, the bandwidth is 50 MHz, and the required carrier-to-noise ratio for reliable operation is 12 dB, find the following:

   (a) The required received signal level.
   (b) The allowable path loss.
   (c) The maximum allowable range.

   Assume that free-space loss applies.

■■■■■■■ **CHAPTER 5**

Radar Systems

5.1 INTRODUCTION

Radar is an acronym for radio detection and ranging. So far the subject has been one-way communication links, with the assumption that the reverse link is identical. While distinct from a classical communication system, radar uses electromagnetic wave propagation and communication technology to perform distance, angle, and/or velocity measurements on targets of interest. Early radar was designed to detect large objects at a distance, such as aircraft or vessels. The first radars were pulse-mode radars using crude angle measurement techniques. As development continued, split antenna apertures were used to sense the direction of the returning wavefront (monopulse radar), providing greater angular accuracy. Another key development was the use of filter banks to detect the Doppler shift on the return signal which provided the operator with target velocity information. Modern imaging radars can also provide information on target size and sometimes target identification by using pulse compression and other range resolution techniques as well as synthetic aperture processing. Contemporary systems also map the earth's surface and provide detailed weather information. A radar "target" may be an aircraft, vehicle, vessel, weather front, or terrain feature, depending upon the application. This chapter provides a general overview of radar operation, primarily from the RF propagation standpoint. For a more in-depth treatment, consult the references.

Consider the radar configuration shown in Figure 5.1, where, the transmitter and receiver are co-located and share one antenna (monostatic radar) and the received signal is a reflection from a distant object. A pulse of RF energy is transmitted and then the receiver listens for the return, similar to sonar. The strength of the return signal depends upon the transmitted power, the distance to the reflecting target, and its electrical size or reflectivity. The radar receiver determines the distance to the target by measuring the time delay between transmission and reception of the reflected pulse. By measuring the phase front (angle of arrival) of the returning signal, the radar

Introduction to RF Propagation, by John S. Seybold
Copyright © 2005 by John Wiley & Sons, Inc.

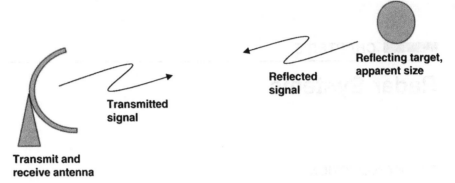

Figure 5.1   Typical radar and target configuration.

can also estimate the location of the target in azimuth and elevation within its beamwidth.

5.2   THE RADAR RANGE EQUATION

The radar range equation is used to determine the received radar signal strength, or signal-to-noise ratio. It can be used to estimate the maximum distance at which a target of a given size can be detected by the radar. The development of the radar range equation is straightforward based on the material from the earlier chapters. The power density at a distance, d, is given by

$$S = \frac{\text{EIRP}}{4\pi d^2} \quad \text{watts/m}^2 \tag{5.1}$$

The power available at the output of an antenna in this power density field is the product of the power density at that point and the antenna's effective area (also called the capture area):

$$P_R = \frac{P_T G_T}{4\pi d^2} \cdot A_e \quad \text{watts} \tag{5.2}$$

Substituting the expression for antenna gain (3.3) yields the Friis free-space loss equation:

$$P_R = \frac{P_T G_T G_R \lambda^2}{(4\pi)^2 d^2} \quad \text{watts} \tag{5.3}$$

Or the path loss can be expressed as

$$L = \frac{P_R}{P_T} = \frac{G_T G_R \lambda^2}{(4\pi d)^2} \tag{5.4}$$

Instead of a receive antenna effective area, in radar, the return signal strength is determined by the *radar cross section* (RCS) of the "target." The RCS is a measure of the electrical or reflective area of a reflector. It may or may not correlate with the physical size of the object. It is a function of the object's shape, composition, orientation, and possibly size. The RCS is expressed in m², or dBsm (10log(RCS in m²)) and the symbol for target RCS is σ_t. Since the RCS is an intrinsic property of the target or reflector, it does not depend on the range or distance from which it is viewed. It may, however, vary with the frequency and the polarization of the radar wave and with the aspect angle from which it is viewed.

When measuring the RCS of a given target, there is oftentimes debate as to whether the average RCS should be computed by averaging the RCS in m² or in dBsm. Averaging in dBsm corresponds to the geometric average of the m² values. Each technique has its advantages and for log-normal RCS distributions, the two are mathematically related [1].

The strength of the reflected signal is determined from the power density at the target by replacing the receive antenna capture area with the RCS.

$$P_{refl} = \frac{P_T G_T}{4\pi d^2} \cdot \sigma_t \qquad \text{watts} \tag{5.5}$$

The power density back at the radar receiver from the reflected signal is

$$S_R = \frac{P_T G_T \sigma_t}{4\pi d^2} \cdot \frac{1}{4\pi d^2} \qquad \text{watts}/m^2 \tag{5.6}$$

When multiplied by the effective area of the radar antenna, this becomes the received power.

$$P_R = \frac{P_T G_T \sigma_t A_e}{(4\pi)^2 d^4} \qquad \text{watts} \tag{5.7}$$

By solving (3.3) for A_e and substituting into (5.7), the expression for the received power can be written in terms of the antenna gains. For a monostatic radar, $G_T = G_R$.

$$P_R = \frac{P_T G_T G_R \sigma_t \lambda^2}{(4\pi)^3 d^4} \qquad \text{watts} \tag{5.8}$$

Using the minimum detectable receive power, one can solve the radar equation for d and find the maximum distance at which detection is possible. In radar, it is customary to use R for range instead of d for distance, so

$$R_{max} = \sqrt[4]{\frac{P_T G^2 \sigma_t \lambda^2}{P_{Rmin}(4\pi)^3}} \quad \text{m} \tag{5.9}$$

This is called the *radar range equation*. It provides an estimate of the maximum radar detection range for a given target RCS. The maximum range is proportional to the transmit power, antenna gain, and target RCS and inversely proportional to the minimum required receive power and the radar frequency.

The RCS as used here is a mean or median value. It is common for the RCS of real targets to fluctuate considerably as the target orientation and the radar-target geometry vary. This fluctuation can considerably degrade the performance of the radar receiver. In fact, modeling of RCS fluctuations and the design of optimal detectors for different RCS statistics is a rich field of study with many books and articles dedicated to the subject [2–9]. Table 5.1 gives approximate expressions for the RCS of several standard calibration targets as a function of dimensions and wavelength. With the exception of the sphere, even these targets have RCS's that vary considerably with aspect angle.

TABLE 5.1 Maximum RCS for Some Calibration Targets from Refs. 10 and 11

Sphere:	$\sigma = \pi r^2$ if $\lambda < \dfrac{r}{1.6}$	For larger wavelengths, the RCS is in either the Mie or the Rayleigh region and more complex expressions apply.
Cylinder	$\sigma = \dfrac{2\pi r h^2}{\lambda}$	Normal (broadside) incidence
Dihedral (diplane):	$\sigma = \dfrac{8\pi h^2 w^2}{\lambda^2}$	
Trihedral (square):	$\sigma = \dfrac{12\pi w^4}{\lambda^2}$	
Flat rectangular plate:	$\sigma = \dfrac{4\pi l^2 w^2}{\lambda^2}$	Normal incidence

Example 5.1. Consider a radar system with the following parameters:

$f = 2\,\text{GHz}$	Transmitter frequency ($\lambda = 0.15\,\text{m}$)
$P_T = 1\,\text{W}$	Peak transmitter power ($P_T = 0\,\text{dBW}$)
$G_T = G_R = 18\,\text{dB}$	Antenna gains
$R = 2\,\text{km}$	Range to target
$B = 50\,\text{kHz}$	Receiver bandwidth
$F = 5\,\text{dB}$	Receiver noise figure
$\sigma_t = 1\,\text{m}^2$	Target radar cross section

What is the SNR at the receiver?

The received signal level is

$$P_R = \frac{P_T G_T G_R \sigma_t \lambda^2}{(4\pi)^3 d^4}$$

This can be computed in dBW as

$$P_R = 0 \text{ dBW} + 18 \text{ dB} + 18 \text{ dB} + 0 \text{ dBsm}$$
$$+ 20\log(0.15) \text{ dBsm} - 30\log(4\pi) - 40\log(2000) \text{ dB-m}^4$$
$$P_R = -145.5 \text{ dBW or} -115.5 \text{ dBm}$$

The receiver noise power is $N = kT_0 BF$ or

$$N_{dBm} = -174 \text{ dBm/Hz} + 10\log(50{,}000) \text{ dB-Hz} + 5 \text{ dB}$$
$$N_{dBm} = -122 \text{ dBm}$$

and the SNR is

$$\text{PR} - N_{dBm} = 6.5 \text{ dB} \quad \square$$

Note that the received power is inversely proportional to R^4, so doubling the distance reduces the signal level by 12 dB. The round-trip path loss is *not* equal to 3 dB (or 6 dB) more than the one-way path loss. It is double the one-way loss in dB (i.e., the round-trip path loss is equal to the one-way path loss squared).

To determine the signal-to-noise ratio at the receiver, the noise-equivalent bandwidth of the receiver must be known. Since conventional pulse radar works by transmitting a short RF pulse and measuring the time delay of the return, the receiver bandwidth will be matched to the transmitted pulse. In this case, the bandwidth of the matched filter receiver is often approximated by $1/\tau$, where τ is the pulse width (this is used as the noise-equivalent bandwidth in noise calculations). The noise floor of the radar receiver can then be estimated by applying (4.9):

$$N = kT_0 \frac{1}{\tau} F$$

The pulse width, τ, also determines the range resolution of the radar. The approximate range resolution of a radar system is given by

$$\Delta r = \frac{c\tau}{2} \tag{5.10}$$

This is approximate because the actual resolution depends on the pulse shape as well as the duration. Shorter pulses require larger receive bandwidths (more noise) and provide less average power (less signal) but better range resolution. The matched filter has an impulse response that matches the transmitted pulse. The range to the target is

$$R = \frac{c \cdot \Delta t}{2} \tag{5.11}$$

where Δt is the elapsed time between transmission and reception of the pulse.

In a pulse or pulse-Doppler radar, the pulses are usually transmitted periodically. This period is called the pulse-repetition interval (PRI) or the pulse-repetition time (PRT). The term pulse repetition period has historically been used as well, but it is not always appropriate since the pulses are not always strictly periodic. The pulse repetition frequency is

$$\text{PRF} \equiv \frac{1}{\text{PRI}}$$

The PRI defines the maximum unambiguous range of the system:

$$R_{unamb} = \frac{c \cdot \text{PRI}}{2} \tag{5.12}$$

A large target beyond the unambiguous range may be interpreted as a closer target. This occurs when the radar is listening for the return of the nth pulse and actually receives a return from the $(n - 1)$th pulse. This is called a multiple-time-around echo, and Figure 5.2 shows a diagram of how this can occur.

For multiple-time-around returns to be an issue, the RCS of the distant reflector must be sufficiently large so that the reflection can overcome the additional path loss and still be detected. Ideally, R_{unamb} should be well beyond

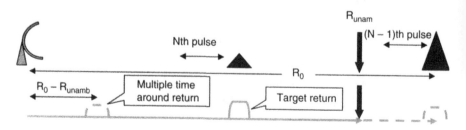

Figure 5.2   Depiction of radar range ambiguities for pulsed radar.

the maximum detection range of the radar so that even large targets will not provide a large enough return to be detected (maximum detection range is defined for a given target RCS). In practice, this is not always practical, but there are ways to mitigate the effect of range ambiguities. One technique is to stagger the PRI's. If the PRI is different from one pulse to the next, the true target returns will remain stationary from one PRI to the next, while the ambiguous returns will appear at a different range for each different PRI and can thereby be eliminated from consideration.

5.3 RADAR MEASUREMENTS

Radar has progressed from a means of detection and ranging to a sensor system that can provide complete position and velocity information on a target as well as potentially imaging the target. The actual measurement can be broken down into range measurement, angle measurement (azimuth and elevation angle), and signature measurement (high-range and cross-range resolution imaging of a target). The basic technology employed for each of these types of measurements is discussed in the sections that follow.

5.3.1 Range Measurement

The most basic method of range measurement using a pulse radar is to simply identify which range gate (receiver time sample) contains the majority of the target energy. This provides a range measurement resolution on the order of the pulse width:

$$\Delta R = c\,\tau/2$$

To improve the range resolution without having to resort to extremely short pulses (which increases the bandwidth required and raises the thermal noise floor), pulse compression may be used. Pulse compression consists of further modulating the radar signal within the transmit pulse. This can be done by chirping the RF frequency or by applying a BPSK modulation with low autocorrelation sidelobes. The result is a substantial increase of the range resolution of the radar without shortening the transmit pulse.

Another approach to enhanced range resolution is using a stepped frequency sequence of pulses, which are then Fourier-transformed to provide a high-range resolution profile of the target [12]. This approach is often used in millimeter-wave radar systems and is frequently used in conjunction with synthetic aperture techniques to provide two-dimensional high-resolution target images.

A less exotic means of improving range measurement resolution is to simply interpolate between adjacent range gates that contain the target energy. A similar approach for tracking radar is called the *split gate tracker*,

Early Gate Late Gate

Range

Target Return

Figure 5.3   Split-gate tracking in range.

where the radar positions a pair of range gates straddling the target [13]. Figure 5.3 shows a diagram of a split gate tracker that is straddling a target return with some range error. By comparing the energy in each range gate and dividing by the sum of the energy in each range gate, an error signal can be generated to determine how much to move the range gates to straddle the target. The error signal is

$$\Delta t = \frac{E_1 - E_2}{E_1 + E_2}\,\tau \tag{5.13}$$

where

E_1 is the received energy in the early gate

E_2 is the received energy in the late gate

τ is the length of each range gate

Δt is the additional time shift required to center the range gates over the target

As mentioned earlier, the pulse width is one factor in determining how much energy actually reaches the target to be reflected back to the radar. The actual transmit energy is the peak power of the transmitter multiplied by the width of the transmitted pulse. The peak envelope power (PEP) is referred to as *peak power* in radar. It is still an average power in the sense that the output signal is squared and averaged over an integer number of periods of the carrier, and it does not mean the maximum instantaneous power within a period of the carrier frequency. The *average power* as used in radar actually refers to the peak power multiplied by the transmit duty cycle. That is, the peak power is averaged over a transmit pulse, and the average power is averaged over the entire pulse repetition interval. The normalized $(R = 1\,\Omega)$ peak power of a pulse radar is given by

$$P_{peak} = \int_0^\tau v^2(t)\,dt$$

where τ is the pulse width and $v(t)$ is the voltage at the radar output. The radar average power is given by

$$P_{ave} = \int_0^T v^2(t)\,dt = \frac{\tau}{T}P_{peak}$$

where T is the PRI, $v(t)$ is the voltage at the radar output, and τ is the pulse width. When the power of a radar is specified, it usually refers to the peak power and not the average power. This is because radar systems often employ many different PRIs and pulse widths.

5.3.2 Doppler Measurement

Target velocity can be measured by estimating the Doppler shift on the return signal. This is accomplished by collecting a series of returns and then applying a discrete Fourier transform to them. For ground-based (stationary) radars, the zero Doppler line is often removed to mitigate the effect of ground clutter. By looking at the Doppler cell where the target appears, the direction and magnitude of the target's radial velocity can be determined. The Doppler resolution is determined by the number of pulse returns used for the Fourier transform. More precisely, it is determined by the total time duration covered by the radar samples. The Doppler extent is limited by the PRF or spacing between pulses. If the target velocity exceeds the maximum of the transform, the Doppler is aliased. This is called velocity or Doppler ambiguity and is similar to the concept of range ambiguity. Some radar systems use an estimate of target velocity based on successive range measurements rather than using Doppler processing. This can be useful if sufficient information for Doppler processing is not available.

5.3.3 Angle Measurement

For angle measurements with precision greater than the antenna beamwidth, various beam-splitting techniques must be used. Sequential lobing consists of squinting the beam slightly above and below and then to the left and right of the tracking point. The return from each position is then used to update the estimated target position in azimuth and elevation. Conical scan uses a continuous circular scan about the track point to develop a tracking error signal. Both of these techniques are subject to tracking error due to noise, target fluctuation, and electronic countermeasures. As a remedy, monopulse was developed. Monopulse uses the result of a single pulse to estimate both the azimuth and elevation error in the antenna position relative to the target. This is accomplished by using a split aperture antenna that actually receives four distinct beams simultaneously. Thus any pulse-to-pulse amplitude variations are eliminated as a source of angle measurement error.

There are different types of monopulse [14, 15], including amplitude and phase monopulse. Standard angle tracking is usually accomplished with amplitude monopulse. For simplicity, consider the amplitude monopulse case in one

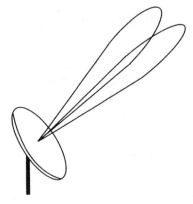

Figure 5.4 Squinted antenna beams for monopulse radar.

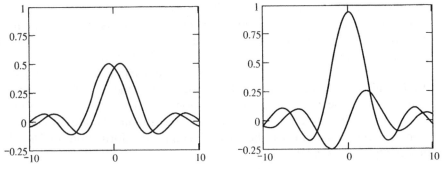

Figure 5.5 Overlapping beams and the corresponding sum and difference beams.

dimension. Two identical beams, squinted in angle, are generated from the same aperture simultaneously by using a specialized feed or phased array. The angle between the two beams is only a fraction of a beamwidth. This is shown pictorially in Figure 5.4, for a single dimension. The signals from the two beams are added and subtracted to produce a sum and a difference pattern. The addition and subtraction can be done in the signal processor, or it may be performed at RF using a hybrid junction such as Magic-Tee [16] or similar device.

When a return signal is received, the samples from the sum and difference patterns are represented by Σ and Δ, respectively. For that reason, the patterns themselves are sometimes referred to as sum and delta patterns. Figure 5.5 shows a representative pair of individual beams and the corresponding sum and difference patterns. The same curves, expressed in dB, are shown in Figure 5.6. This is the conventional way of expressing the sum and difference patterns. The angle of the target from bore-sight is given by the ratio of delta to sum. Both values are complex, so the denominator is rationalized.

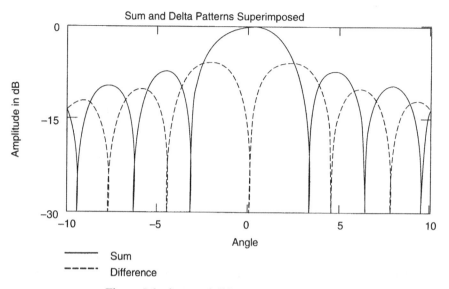

Figure 5.6   Sum and difference patterns in dB.

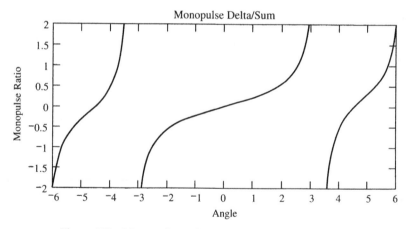

Figure 5.7   Monopulse ratio for the theoretical antenna.

$$\varepsilon = k\frac{\Delta\Sigma^*}{\Sigma\Sigma^*} \tag{5.14}$$

where k is a calibration constant that depends upon the actual beam shapes and angular separation of the two beams. Of course the actual values must be used in this expression, not the dB values. Figure 5.7 shows what the monopulse ratio looks like for this theoretical antenna. For the plot, the value of k is unity. The actual value is chosen to make the slope of the linear part of the curve unity so that angle error is read out directly. It is interesting to note that if enough error is present, the monopulse ratio has other angles where the error slope is positive. It is possible for a radar to see and actually track a target in these regions if it is strong enough to be detected in the sidelobe. This is called sidelobe lock or sidelobe tracking. Tactical systems use various techniques such as a wide-angle spotting antenna to preclude sidelobe lock.

As stated earlier, monopulse angle measurements are very robust to signal amplitude fluctuations because all of the data comes from the same pulse. The effect of noise on the monopulse ratio is not as well behaved, however. A theoretical analysis shows that the variance of the monopulse ratio is infinite in the presence of Gaussian noise [17–19]. This is due to the fact that there is a finite probability of Σ being arbitrarily close to zero. This fact is not of great practical concern because measurements that are that weak will be discarded. It does, however, accurately suggest that monopulse is susceptible to jamming by white noise. For that reason, regular monopulse cannot be used to track a noise source. Instead, the two beams must be processed separately, and the detected power in each must be used to determine the angle tracking error.

5.3.4 Signature Measurement

Many modern radar systems employ radar signature measurement techniques to permit the radar to identify targets. The most basic type of signature measurement is the high-range resolution profile. If multiple profiles are coherently collected and stored with regular azimuthal spacing, it is possible to generate a cross-range profile for each cell and the range profile [20]. This produces a two-dimensional map of the target. The results can be used for target identification, system diagnostics, terrain mapping, and other remote sensing applications. The azimuthal spacing is ideally in angle, with the radar moving in an arc whose focus is at the target, or a stationary radar viewing a target rotating at a constant angular velocity. In many applications, the motion is lateral instead of angular, but it is possible to correct for the difference and still generate very high quality images.

The preceding technique is called *synthetic aperture radar* (SAR) if the target is stationary or *inverse synthetic aperture radar* (ISAR) if the radar is static and target rotation is used to produce the required changes in perspective. The images from SAR or ISAR processing are range/azimuth images. The dimension of height can also be added by processing a set of coherent range-azimuth maps taken at different elevation perspectives [21]. This method is principally used for instrumentation purposes because the required setup, calibration, and data processing preclude most other applications. This is called three-dimensional ISAR. Some radar systems also perform a monopulse measurement on the contents of each resolution cell in a range-azimuth map to provide limited information on the height of the scatter(s) in each cell. This is also sometimes called three-dimensional ISAR, but that is a misnomer since there is not actually any resolution in elevation, but rather a single composite value.

5.4 CLUTTER

Clutter is defined as any unwanted radar echo. For a weather radar system, an aircraft might be considered clutter; similarly, for a terrain mapping radar, a vehicle in the field of view would be considered clutter. For the purposes of this chapter, it is assumed that a vehicle or aircraft is the desired target and that the natural reflectors in the scene represent clutter. Radar returns from ground targets will include reflected energy from ground clutter—that is, the surrounding terrain and vegetation. This type of clutter is called *area clutter* since it is predominantly two-dimensional in nature. The amount of clutter depends on the resolution of the radar in range and angle and the terrain characteristics, as will be shown.

Airborne target returns seen by upward-looking radar may include *volume clutter* from moisture or hydrometeors in the propagation path. Another source of volume clutter is intentional clutter such as chaff (small metallic strips) dispersed by an aircraft to overwhelm the radar receiver and conceal the presence of aircraft. Chaff is an example of a radar countermeasure, but it is treated as clutter since the effect is the same. Volume clutter as its name implies is three-dimensional in nature. An upward looking radar may also experience ground (area) clutter returns entering through the antenna sidelobes. For moderately powered, ground-based radar, this can be a significant source of (stationary) clutter.

In addition to estimating the mean clutter return signal, the radar designer must take into account the statistics of the clutter. Some clutter such as uniform terrain or calm water may have very little variation, whereas rough terrain may have considerable spatial variation and tall vegetation or bodies of water in heavy wind will also show significant temporal variation.

When designing a radar system for operation in a clutter environment, the signal-to-clutter ratio (SCR) and the clutter-to-noise ratio (CNR) are of interest. The SCR provides a measure of the expected target signal strength versus the expected clutter return strength, which is used for detection analysis and prediction. The CNR provides an indication of whether target detection is limited by noise or limited by the external clutter.

5.4.1 Area Clutter

Area clutter, which is synonymous with ground clutter, is characterized by the average clutter cross section per unit area, σ^0 (sigma-zero). This parameter is called the *backscatter coefficient*, and the units are m^2/m^2. The amount of clutter received depends upon the backscatter coefficient and how much of the ground area is illuminated. The backscatter coefficient of clutter depends on the grazing angle.

If the extent of the ground clutter is larger than the radar footprint, then the width of the clutter patch is defined by the antenna azimuth beamwidth and the range or distance to the clutter patch. The (down-range) length of the clutter patch is determined by either the range gate size for shallow grazing angles, or by the elevation beamwidth for steeper grazing angles [22].

The width of the clutter cell is

$$W \cong R\theta_{AZ} \tag{5.15}$$

where R is the range to the center of the clutter cell and θ_{AZ} is the two-way antenna azimuth beamwidth in radians. The two-way beamwidth is used because the same beam is illuminating and then receiving the reflection from the clutter. Another definition of the two-way beamwidth is the -1.5-dB beamwidth of the antenna pattern. The concept of the width of the ground clutter cell is illustrated in Figure 5.8. The depth or length of the clutter cell is determined by the smaller of the range gate projected onto the ground, or the elevation beamwidth times the total range projected onto the ground.

Thus the length of the clutter cell can be expressed as

$$l \approx \min\left(\frac{c\tau}{2}\sec(\psi), R\theta_{EL}\csc(\psi)\right) \tag{5.16}$$

Figure 5.9 shows the geometry for the low grazing angle case where the range gate size determines the down-range dimension of the clutter area. At higher grazing angles, the antenna elevation pattern becomes the limiting factor as illustrated in Figure 5.10. For the high grazing angle case, the two-way elevation beamwidth must be used rather than the one-way beamwidth, just as was done in the range-gate-limited case with the azimuth beamwidth.

So at shallow grazing angles, where the clutter cell size is pulse-limited, Figure 5.11 applies and

$$A_c \cong R\theta_{AZ}\frac{c\tau}{2}\sec(\psi) \tag{5.17}$$

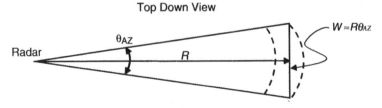

Top Down View

Figure 5.8   Determination of total illuminated clutter area.

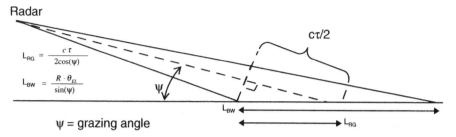

Figure 5.9   Range-gate-limited clutter area.

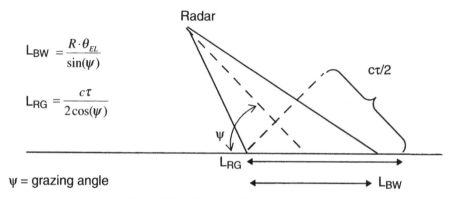

Figure 5.10   Beam-limited clutter area.

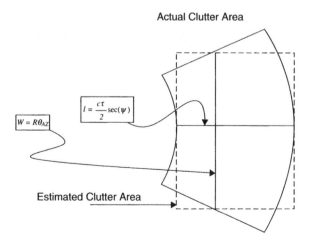

Figure 5.11   Shape of range-gate-limited clutter cell.

At steeper angles, when beam-limited, the clutter area is simply the area of the elliptical footprint as shown in Figure 5.12:

$$A_c = \frac{\pi}{4} R^2 \theta_{EL} \theta_{AZ} \csc(\psi) \tag{5.18}$$

The total radar cross section of the clutter in the radar cell is the clutter patch area multiplied by the backscatter coefficient:

$$\sigma_c = \sigma^0 A_c \tag{5.19}$$

Consider the pulse-limited, low grazing angle case where (5.17) applies. From (5.8), the clutter reflection power seen at the radar is

$$P_c = \frac{P_T G^2 \lambda^2 \sigma_c}{(4\pi)^3 R^4}$$

Substituting the expression for the σ_c into (5.8) yields

$$P_c = \frac{P_T G^2 \lambda^2 \sigma^0 \theta_{AZ} \frac{c\tau}{2} \sec(\psi)}{(4\pi)^3 R^4}$$

or

$$P_c = \frac{P_T G^2 \lambda^2 \sigma^0 \theta_{AZ} \frac{c\tau}{2} \sec(\psi)}{(4\pi)^3 R^3} \tag{5.20}$$

Thus as the distance, R, increases, the clutter power decreases as $1/R^3$, whereas the target power decreases as $1/R^4$. This is due to the beam spreading with dis-

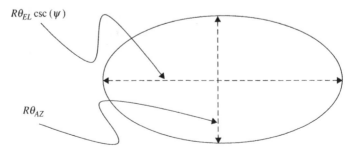

Figure 5.12 Shape of beam-limited clutter cell.

tance increasing the width of the radar footprint in the clutter patch and therefore the amount of clutter return that is seen. For the beam-limited case, the clutter area is a function of R^2 rather than R, so the return clutter power is proportional to $1/R^2$, versus $1/R^4$ for the target power.

As seen from the preceding development, these values are approximate. In addition, the antenna beam has a roll-off, and it does not drop off to zero gain at the edge of the 3 dB (or 1.5 dB) beamwidth. Even so, this is standard practice for determining the expected clutter return and it provides good estimates. The limiting factor on predicting clutter return is most often the variability in the clutter backscatter coefficient of the actual clutter encountered.

There are a variety of ways that the system designer can mitigate the effects of clutter. These methods include (a) using narrow antenna beams and short pulses to reduce the amount of clutter return that is seen (b) averaging multiple "looks" when the clutter has a short correlation time such as vegetation or (c) making use of Doppler. If the target is moving, it is possible to take multiple looks and then filter the sequence (FFT) to extract the moving target. If the radar is stationary, the clutter will be centered at the zero Doppler bin and can be filtered out without losing a moving target.

If the radar is airborne, the ground clutter will have a nonzero Doppler shift, but it can be identified and ignored as long as the target has motion relative to the ground. The number of range samples used and their spacing determine the Doppler resolution. There can be Doppler ambiguities if the Doppler shift is sufficiently large. This is a form of aliasing and is controlled by decreasing the time between samples, PRI, at the expense of the unambiguous range [23].

Example 5.2. Consider a radar system with

$$P_T = 1,000,000 \text{ W}$$
$$G_{ANT} = 28 \text{ dB}$$
$$\tau = 100 \text{ μs}$$
$$T_{eff} = 200 \text{ K}$$
$$f = 10 \text{ GHz}$$

What is the SNR of the return from a 1-m² target at 20 km?
The wavelength is

$$\lambda = c/f = 0.03 \text{ m}$$

and the bandwidth is approximately

$$B \approx 1/\tau = 10 \text{ kHz}$$

To compute the noise floor, the noise figure must first be determined:

$$F = 10 \log(1 + T_{\text{eff}}/T_0) = 2.3 \text{ dB}$$

Then the total noise power in the bandwidth of interest can be computed:

$$N = -204 \text{ dBW/Hz} + 10 \log(10 \text{ kHz}) + F$$

$$N = -161.7 \text{ dBW}$$

Next, compute the return or received signal power

$$P = \frac{P_T G_T G_R \sigma \lambda^2}{(4\pi)^3 R^4}$$

which yields

$$P = -119.5 \text{ dBW}$$

Finally, the signal-to-noise ratio is

$$\text{SNR} = 42.2 \text{ dB} \quad \square$$

Example 5.3. For the same system as Example 5.2, if the azimuth beamwidth is $\theta_{AZ} = 0.3°$, the grazing angle is $\psi = 5°$, and $\sigma^0 = 0.01$, what is the signal-to-clutter ratio (SCR) and the clutter-to-noise ratio (CNR)?

First the clutter return is computed. The clutter area is given by (5.17):

$$A_c = R\theta_{AZ} \frac{c\tau}{2} \sec(\psi)$$

So,

$$A_c = 1{,}575{,}000 \text{ m}^2$$

which can be expressed as 62 dB-m^2 or 62 dBsm.

Next compute the power received from clutter reflection:

$$P_c = \frac{P_T G^2 \lambda^2 \sigma^0 A_c}{(4\pi)^3 R^4}$$

So the clutter return is

$$P_c = -77.5 \text{ dBW}$$

The signal-to-clutter ratio is

$$\text{SCR} = -42 \text{ dB}$$

and the clutter-to-noise ratio is

$$CNR = 84.2 \text{ dB}$$

So for this system, clutter is the limiting factor (80 dB over the noise) and the signal is buried in clutter. It would require a significant amount of processing to extract a meaningful signal from 42 dB below the clutter. ☐

5.4.2 Volume Clutter

When considering volume clutter, the amount of clutter that is illuminated depends upon the range gate length, the range to the range gate of interest, and the azimuth and elevation beamwidths of the antenna as depicted in Figure 5.13. The volume of the clutter cell will be approximately [24]

$$V = (\pi/4)(c\tau/2)R^2\theta_{EL}R\theta_{AZ} \qquad \text{m}^3 \tag{5.21}$$

where

c is the velocity of propagation
τ is the pulse width
R is the range to the center of the clutter cell
θ_{EL} and θ_{AZ} are the two-way, 3-dB, elevation and azimuth antenna beamwidths (or the 1.5-dB beamwidths)

Volume clutter is characterized by the backscatter cross section per unit volume, η, which has units of m^2/m^3. The total clutter cross section then becomes

$$\sigma = \eta\frac{\pi}{4}\frac{c\tau}{2}R^2\theta_{AZ}\theta_{EL} \qquad \text{m}^2 \tag{5.22}$$

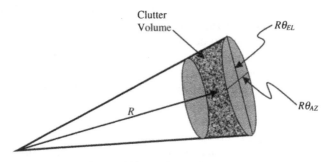

Figure 5.13 Volume clutter cell.

So when the radar range equation is applied, the clutter return only decreases as R^2 because of the dependence of the clutter volume on R^2.

The backscatter cross section per unit volume can be approximated by [25]

$$\eta = 7f^4 RR^{1.6} \times 10^{-12} \text{ m}^2/\text{m}^3 \tag{5.23}$$

where f is the frequency in GHz and RR is the rain rate in mm/hr.

Other potential sources of volume clutter include snow and hail, which are difficult to characterize because of their wide variability and limited data. Dust, fog, and smoke are significant attenuators in the visible light band, but have little or no attenuation or backscatter effect on radar and are generally ignored.

5.4.3 Clutter Statistics

The statistics of clutter that must be considered are the temporal and spatial variability. Moving clutter such as seawater, moving foliage, or blowing rain and snow present a time-varying clutter background to a radar system. This can result in significant false alarms or missed detects if not designed for. Similarly, static clutter will vary from location to location and over look angle (both azimuth and elevation). Again, this variability must be considered by the radar designer if successful performance is to be achieved. Various types of detection algorithms, range coding Doppler processing, and other techniques are used to mitigate the effects of clutter.

5.5 ATMOSPHERIC IMPAIRMENTS

The atmospheric impairments that affect communication links also affect radar systems. The effect is compounded for radar systems since it entails two-way propagation, plus the potential for backscatter since the receiver is tuned to same frequency that was just transmitted. The backscatter effects are treated in the previous section on volume clutter, so only the attenuation effects are discussed in the sections that follow.

Multipath is a significant problem for many radar installations. The reception of more than one return from the target can cause significant fading; more importantly, however, it disrupts the phase front of the received wave, which reduces the radar's ability to accurately measure the target angle. Multipath can also have the undesired effect of spreading the target return in range and thereby corrupting the range measurement. This is mitigated in systems that expect to experience multipath by using leading-edge trackers rather than tracking on the target range centroid.

Atmospheric scintillation can be a concern for radar systems that have elevation look angles near zero degrees (horizontal). The variation due to scintillation is sufficiently slow that it can generally be treated more as a fade in a radar system. The fades can be significant, however, and should be planned for, especially in coastal areas or over large bodies of water. Atmospheric scintillation is usually only a concern during nighttime hours. A related phenomenon is atmospheric ducting, which can be very troublesome for radar systems. When ducting occurs, the propagation loss is not inversely proportional to R^4, but is considerably less. This can cause target detections well beyond the normal range of the radar and in fact sometimes well beyond the radio horizon. Atmospheric scintillation is treated in the following chapter. The models in Chapter 6 are for one-way propagation, however, so appropriate adjustments must be made.

Refraction is the bending of radio waves downward. This is due to the fact that the atmosphere is less dense as elevation increases. The effect is negligible for elevation look angles above 2 or 3 degrees [26]. For low elevation angles, the effect is to extend the radio horizon. The effect is taken into account by using the so-called 4/3's-earth model. The model consists of simply replacing the earth's radius by 4/3 of its actual value in horizon calculations. The 4/3's factor is based on an ITU reference standard atmosphere at sea level and as such represents an average. While not universal, the 4/3's-earth model is very widely used.

The effect of atmospheric loss (absorption) is identical to the absorption for communication systems except that of course it must be doubled to account for the two-way path. A model for atmospheric absorption is presented in the following chapter. Precipitation is covered in Chapter 10 (rain attenuation), so no further discussion is presented here. While more of an environmental than atmospheric impairment, wildlife such as birds or swarms of insects can cause false detections and degrade the performance of some radars.

5.6 SUMMARY

Since radar relies on the reflection of a transmitted signal, radar systems suffer path loss that is proportional to R^4 rather than the R^2 for the one-way propagation of a communication system. The radar range equation is based on the radar cross section and two-way application of the Friis free-space loss equation and provides a means of predicting return signal strength. It can also be used to determine the maximum detection range for a given target RCS. Radar cross section is measured in units of m^2, or dBsm, and may or may not correlate with the physical size.

Radar systems measure range and usually azimuth and elevation angle of targets. Radar may also measure Doppler shift (velocity), the range profile, or range and cross-range images of targets to permit identification or classification of targets. Angular measurement is usually accomplished by using a split aperture antenna and monopulse processing. Range tracking can be performed using two time samples or range gates that straddle the target.

In addition to thermal noise, radar systems must often cope with unwanted environmental reflections called *clutter*. Clutter can be either area (ground) or volume (weather). Since the clutter area or volume grows with R or R^2, the clutter does not decrease with distance as quickly as the signal strength from a limited extent target does. Area clutter is characterized by the backscatter coefficient σ^0, which is the RCS per unit area. The backscatter coefficient is a function of the terrain and the grazing angle of the radar signal. Volume clutter is characterized by the clutter backscatter coefficient per unit volume, usually designated by η and expressed in m^2/m^3.

For those interested in further study on radar systems, the books by Skolnik [23], Stimson [27], and Peebles [10] are a suggested starting point.

REFERENCES

1. J. S. Seybold and K. L. Weeks, Arithmetic versus geometric mean of target radar cross section", *Microwave and Optical Technology Letters*, April 5, 1996, pp. 265–270.
2. A. K. Bhattacharyya and D. L. Sengupta, *Radar Cross Section Analysis & Control*, Artech House, Norwood, MA, 1991.
3. L. V. Blake, *Radar Range Performance Analysis*, Munro Publishing Company, Silver Spring, MD, 1991.
4. J. V. DiFranco and W. L. Rubin, *Radar Detection*, Artech House, Norwood, MA 1980.
5. P. Swerling and W. L. Peterman, Impact of target RCS fluctuations on radar measurement accuracy, *IEEE Transactions on Aerospace and Electronic Systems*, Vol 26, No. 4, July 1990, pp. 685–686.
6. P. Swerling, Radar probability of detection for some additional fluctuating target cases, *IEEE Transactions on Aerospace and Electronic Systems*, Vol 33, No. 2, April 1997, pp. 698–709.
7. P. Swerling, More on detection of fluctuating targets (corresp.), *IEEE Transactions Information Theory*, Vol 11, No. 3, July 1965, pp. 459–460.
8. P. Swerling, Detection of radar echos in noise revisited, *IEEE Transactions Information Theory*, Vol 12, No. 3, July 1966, pp. 348–361.
9. P. Swerling, Probability of detection for fluctuating targets, *IEEE Transactions Information Theory*, Vol 6, No. 2, April 1960, pp. 269–308.
10. P. Z. Peebles, Jr., *Radar Principles*, John Wiley & Sons, New York, 1998, pp. 197–204.
11. N. C. Currie and C. E. Brown, *Principles and Applications of Millimeter-Wave Radar,* Artech House, Norwood, MA 1987, Figure 17.5, p. 769.
12. D. R. Whener, *High Resolution Radar*, Artech House, Norwood, MA, 1987, Chapter 5.
13. M. I. Skolnik, *Introduction to Radar Systems*, McGraw-Hill, New York, NY 2001, pp. 246–248.
14. S. M. Sherman, *Monopulse Principles and Techniques*, Artech House, Norwood, MA, 1984.

15. A. I. Leonov and K. I. Fomichev, *Monopulse Radar*, Artech House, Norwood, MA, 1986.
16. M. I. Skolnik, *Introduction to Radar Systems*, McGraw-Hill, New York, 2001, p. 216.
17. S. M. Sherman, *Monopulse Principles and Techniques*, Artech House, Norwood, MA, 1984, p. 305.
18. I. Kanter, The probability density function of the monopulse ratio for *n* looks at a combination of constant and Rayleigh targets, *IEEE Transactions on Information Theory*, Vol. 23, No. 5, September 1977, pp. 643–648.
19. A. D. Seifer, Monopulse-radar angle tracking in noise or noise jamming, *IEEE Transactions on Aerospace and Electronic Systems*, Vol. 28, No. 3, October 1994, pp. 308–314.
20. D. R. Whener, *High Resolution Radar*, Artech House, Norwood, MA, 1987, Chapter 6.
21. J. S. Seybold and S. J. Bishop, "Three-dimensional ISAR imaging using a conventional high-range resolution radar", 1996 *IEEE National Radar Conference Proceedings*, May 1996, pp. 309–314.
22. F. E. Nathanson, *Radar Design Principles*, McGraw-Hill, New York, 1969, pp. 63–67.
23. M. I. Skolnik, *Introduction to Radar Systems*, McGraw-Hill, New York, 2001, Chapter 3.
24. N. C. Currie, R. D. Hayes, and R. N. Trebits, *Millimeter-Wave Radar Clutter*, Artech House, Norwood, MA, 1992, p. 9.
25. M. I. Skolnik, *Introduction to Radar Systems*, McGraw-Hill, New York, 2001, p. 444.
26. M. I. Skolnik, *Introduction to Radar Systems*, McGraw-Hill, New York, 2001, pp. 499–500.
27. G. W. Stimson, *Introduction to Airborne Radar*, Sci-Tech Publishing, Mendham, NJ, 1998.

EXERCISES

1. Consider a radar system with the following parameters:

$P_T = 100\,\text{W}$
$f = 10\,\text{GHz}$
$T_{eff} = 100\,\text{K}$
$G_{ant} = 28\,\text{dB}$
Pulsewidth $\tau = 1\,\mu\text{s}$
Minimum SNR for reliable detection $= 3\,\text{dB}$

What is the maximum distance at which a 2-m^2 target can be detected?

2. Consider a pulse radar system with the following parameters:

Transmit power	$P_t = 10.0\,\text{kW}$
Antenna gain	$G = 25\,\text{dB}$
Pulse duration	$\tau = 1\,\text{ms}$

$$\text{Receiver noise figure} \quad F = 6\,\text{dB}$$
$$\text{Frequency} \qquad\qquad f = 10\,\text{GHz}$$

What is the minimum target RCS that can be detected at 8 km if the SNR required for detection is 3 dB? Show your work and give your answer in dBsm and m^2.

3. For the same radar in problem 1, if the antenna beamwidth is 3 degrees in each dimension and clutter backscatter coefficient is unity ($\sigma^0 = 0\,\text{dB}$), what is the signal-to-clutter and clutter-to-noise ratio for the system? Assume a grazing angle of 5 degrees.

4. Consider a radar system with the following parameters:

$$\text{Transmit power} \quad P_t = 10.0\,\text{kW}$$
$$\text{Antenna gain} \qquad G = 25\,\text{dB}$$
$$\text{Frequency} \qquad\quad f = 10\,\text{GHz}$$

What is the received signal level from a 13-dBsm target at 10 km?

5. Some radar systems use coherent integration (pre-detection) to improve the signal-to-noise ratio and thereby the sensitivity of a radar system. The principle is that multiple returns from the same target will add coherently, whereas the noise with each return is independent and therefore adds non-coherently. Thus the noise increases as the square root of N and the signal increases as N, where N is the number of pulses or returns that are integrated.

   (a) For the radar given in problem 2, if the same value for minimum RCS is used, but the required signal-to-noise ratio for detection is 7 dB, how many pulses should be integrated assure reliable detection?

   (b) Would the integration have any effect on the signal-to-clutter ratio? Explain.

   (c) Would the integration have any effect on the clutter-to-noise ratio? Explain.

CHAPTER 6

Atmospheric Effects

6.1 INTRODUCTION

Many communication systems rely on RF propagation through the atmosphere. Thus modeling atmospheric effects on RF propagation is an important element of system design and performance prediction. For the purpose of propagation modeling, the atmosphere can be divided into three regions: *troposphere*, *stratosphere*, and *tropopause*. The *troposphere* is the lowest region of the atmosphere where temperature tends to decrease with height and where weather occurs. The *stratosphere* is the region above the weather where air temperature remains constant with height. The *tropopause* is the boundary region between the troposphere and the stratosphere. Some texts also include the effects of the ionosphere as an atmospheric effect. Ionospheric refraction/reflection is limited to frequencies below a few hundred megahertz. There are a few other ionospheric effects, which occur at higher frequencies, such as group delay, Faraday rotation, and fading/scintillation. These effects of the ionosphere are all discussed in Chapter 11, "Satellite Communications," and are not addressed further in this chapter.

The atmospheric effects of interest for RF propagation are refraction/reflection, scattering, and absorption/attenuation. With the exception of refraction, these effects are all minimal below 30 MHz. Between 30 MHz and 1 GHz, refraction/reflection is the primary concern. Above 1 GHz or so, attenuation starts to be a significant factor and refraction/reflection becomes less of an issue except for nearly horizontal paths. Atmospheric multipath also starts to be observed above 1 GHz and can cause extreme fading on terrestrial microwave links. Absorption due to atmospheric gases (including water vapor) is discussed in detail in the sections that follow. The effects of precipitation are also examined in this chapter and in greater detail in Chapter 10 (rain attenuation). The effects of interest for propagation analysis are in the troposphere and, to a limited extent, the tropopause. The stratosphere is well approximated as free-space.

Introduction to RF Propagation, by John S. Seybold
Copyright © 2005 by John Wiley & Sons, Inc.

6.2 ATMOSPHERIC REFRACTION

Refractive and scattering effects of the atmosphere include:

- Refraction on horizontal paths resulting in alteration of the radio horizon due to ray curvature.
- *Troposcatter*, from localized fluctuations in the atmospheric refractive index, which can scatter electromagnetic waves.
- *Temperature inversion*, abrupt changes in the refractive index with height causing reflection.
- *Ducting*, where the refractive index is such that electromagnetic waves tend to follow the curvature of the earth.

These effects vary widely with altitude, geographic location, and weather conditions. The effects can permit beyond-the-horizon communication (or interference), or produce blockage and diffraction from terrain that appears to be below the line of sight and multipath fading. These effects are discussed in the sections that follow.

6.2.1 The Radio Horizon

Gradual changes in the refractive index with height cause EM waves to bend in the atmosphere. If the atmosphere were homogeneous, the waves (rays) would travel in a straight line and the physical and RF horizons would coincide. The rate of change of the refractive index with height can be approximated as being constant in the first kilometer above sea level [1]. This increases the apparent distance to the horizon, by bending the nearly horizontal rays downward. This is illustrated in Figure 6.1, where a horizontal ray is bent downward and sees a radio horizon that is beyond the straight line-of-sight horizon (*h* is the antenna height). By replacing the model of the earth's surface with one that has a radius of 4/3 of the earth's radius (for a standard reference atmosphere), the curved ray becomes straight. This is shown in Figure 6.2 where the horizon appears at the point where a straight ray emanating from

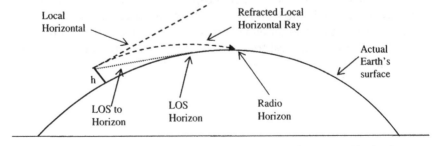

Figure 6.1 Effect of atmospheric refraction on the distance to the horizon.

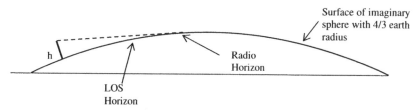

Figure 6.2   Equivalent radio horizon with 4/3 earth radius model.

the source intersects the 4/3 R_e spherical surface. The 4/3 earth radius approximation is derived based on a standard atmosphere at sea level and is therefore not universally applicable. Nonetheless, it is very widely used and often treated as if it were universal.

It is interesting to note that the ray that reaches the radio horizon is not the same ray that is directed at the LOS horizon, it leaves the source at a slightly greater elevation angle. In the case of very narrow beamwidth antennas, it is possible that the departure angle of the radio horizon ray is outside of the beamwidth of the antenna. If this occurs, then the effect of the refraction can be a reduction in communication distance rather than an increase since the ray leaving the source at the center of the antenna beamwidth will be bent toward the ground and the peak of the transmitted signal will not be directed at the intended receiver.

The energy that reaches the receiver will be a combination of off-axis radiation from the antenna and any diffraction from intervening terrain that blocks the bore-sight radiation. For this reason, it is important to verify antenna aiming at different times over a period of days [2, p. 20]. If aiming is only done once and happens to be done during a period of significant refraction, then under nominal conditions the link margin will be below the design level, reducing the link margin and availability.

6.2.2 Equivalent Earth Radius

The radius of the earth and the rate of change of the refractive index can be used to determine the equivalent radius of the earth based on local conditions [3]:

$$r_{eq} = \frac{1}{\dfrac{1}{r} + \dfrac{dn}{dh}} \tag{6.1}$$

The refractive index can be expressed as

$$n = \left(1 + N \times 10^{-6}\right) \tag{6.2}$$

where N is called the *refractivity*:

$$N = (n-1) \times 10^6 \qquad (6.3)$$

From ITU-R P.453-8 [1], N can be modeled as

$$N = \frac{77.6}{T}\left(P + \frac{4810e}{T}\right) \qquad \text{N units} \qquad (6.4)$$

where

   P is the total pressure
   e is the water vapor pressure
   T is the absolute temperature

Using the typical values for a standard atmosphere [4] yields $N = 312$. The typical values are

   $P = 1000\,\text{mb}$
   $e = 10\,\text{mb}$
   $T = 290\,\text{K}$

Thus the refractivity of a standard atmosphere is

$$N_S = 312$$

The refractive index is approximately unity at sea level and drops off nearly exponentially with height. The refractivity as a function of height can be modeled as [1, 5]

$$N = N_S e^{-h/H} \qquad (6.5)$$

where $H = 7\,\text{km}$. From (6.2), it can be seen that

$$\frac{dn}{dh} = \frac{dN}{dh} \times 10^{-6}\,\text{N units/km} \qquad (6.6)$$

and from (6.5),

$$\frac{dN}{dh} = \frac{-N_S}{H} e^{-h/H} \qquad (6.7)$$

For altitudes below 1 km, dN/dh is well-approximated by its value at 1 km:

$$\frac{dN}{dh}(1\,\text{km}) = \frac{-312}{7} e^{-1/7} = -38.6\,\text{N units/km} \qquad (6.8)$$

Therefore when $h < 1\,km$, the expression for the refractivity gradient is

$$\frac{dN}{dh} \approx -39\ \text{N units}/\text{km} \qquad (6.9)$$

Using the preceding expression, one can determine the required adjustment to make the local horizontal ray straight:

$$\frac{1}{r_{eq}} = \frac{1}{r} + \frac{dn}{dh} \qquad (6.10)$$

Thus

$$\frac{r}{r_{eq}} = \frac{1}{k} = 1 + r\frac{dn}{dh} \qquad (6.11)$$

where r is the earth's radius (6370 km) and

$$\frac{dn}{dh} = -39\ \text{N units}/\text{km} = -39 \times 10^{-6}\,\text{km}^{-1}$$

So

$$k = \frac{1}{1 - 6370 \times 39 \times 10^{-6}} = \frac{1}{0.75} = \frac{4}{3}$$

as expected for the standard atmosphere conditions that were used.

Example 6.1. Find the distance to the radio horizon from a 50-m tower installed at 2 km above sea level, given the following atmospheric conditions at sea level:

$P = 1100\,mb$
$e = 12\,mb$
$T = 260\,K$

The expression for the refractivity is obtained by applying (6.4) with the precoding parameter values,

$$N = \frac{77.6}{T}\left(P + \frac{4810e}{T}\right) \qquad \text{N units}$$

yields

$$N_S = 394.57 \, \text{N units}$$

Using the expression for the refractivity gradient and the 2-km height yields

$$\frac{dN}{dh}(h) = \frac{-N_S}{7} e^{-h/7}$$

Thus

$$\frac{dN}{dh} = -42.36$$

and

$$\frac{dn}{dh} = -0.00004236$$

The value of k can be found from

$$\frac{r}{r_{eq}} = \frac{1}{k} = 1 + r \frac{dn}{dh}$$

Using

$$r = 6370 \, \text{km}$$

for the actual earth's radius yields

$$k = 1.370$$

The distance to the horizon is then given by

$$d \cong \sqrt{2krh_T}$$

The distance to the horizon is

$$d = 29.5 \, \text{km} \quad \square$$

6.2.3 Ducting

Ducting occurs when the curvature of the refraction is identical (or nearly identical) to the curvature of the earth. Another way to look at it is that under ducting conditions, the equivalent earth radius is infinite. There are two primary types of ducts: surface ducts, where the wave is trapped between the earth's surface and an inversion layer, and elevated ducts, where the layers of the atmopshere trap the wave from both above and below. Elevated ducts may occur up to about 1500 m above the surface. In either case, both ends of the communication link must be located within the same duct for a signal enhancement to be observed. The case of surface ducting is more common and usually of much greater practical concern. According to Parsons [6], ducting tends to be primarily an effect due to water vapor content, but it can also be caused by a mass of warm air moving in over cooler ground or the sea or by night frost.

The radius of curvature of an electromagnetic wave can be written as

$$\rho = -\frac{dh}{dn}$$

If ρ is equal to the radius of the earth, then the electromagnetic wave will follow the curvature of the earth:

$$\frac{1}{r_{eq}} = \frac{1}{r} + \frac{dn}{dh}$$

This case corresponds to $r_{eq} = \infty$, or

$$\frac{1}{r} + \frac{dn}{dh} = 0$$

Therefore, an EM wave will follow the curvature of the earth when

$$\frac{dn}{dh} = -\frac{1}{r} = -\frac{1}{6370} = -157 \times 10^{-6}\,\text{km}^{-1}$$

This creates a surface duct condition where RF can propagate well beyond the physical horizon. The path loss can be considerably less than free space, analogous to a waveguide. A duct may only be 1–2 m high, resulting in a significant concentration of the wave. This can be a limitation for frequency reuse (interference), but it also serves to increase the range of terrestrial line-of-sight communications.

6.2.4 Atmospheric Multipath

Signal enhancement is not the only effect of refraction or ducting. When sufficient refraction occurs, it can appear like a reflection, similar to ionospheric propagation. If this occurs while there is still a direct line-of-sight path, or if several paths exist, the resulting atmospheric multipath can create signal enhancement and/or signal attenuation. Fades due to atmospheric multipath are a significant consideration, particularly for point-to-point microwave links. The effect occurs predominantly in higher-humidity areas during nighttime hours, with coastal areas being particularly susceptible. Like refraction, atmospheric multipath only affects paths that are very nearly horizontal. Atmospheric multipath is primarily observed over very flat terrain; irregular terrain makes formation of a uniform atmospheric layer unlikely.

The impact of this kind of multipath on terrestrial point-to-point microwave links was studied by Bell Laboratories in the 1960s and 1970s. Barnett [7] produced a model to characterize the observed multipath effects, which has been widely used in the United States. About the same time, Morita [8] developed a model based on measurements made in Japan. Later, Olson and Segal [9] performed a similar study in Canada and developed a model for the observed effects in Canada. A recent article by Olsen et al. [10] compared several models to measured data and concluded that the ITU model [11] slightly outperformed the other models. Application of the latest ITU model [2] is presented here.

The ITU model for atmospheric multipath has two different formulations for low probability of fade and another formulation for all fade probabilities. For most applications the lower fade probability are suitable. In addition to providing multipath fade depth predictions, the ITU also provides a model for multipath signal enhancement. The enhancement model is not presented herein, but it may be of interest in assessing the potential for interference in frequency re-use applications.

The first step in applying the ITU model for small percentages is to determine the appropriate geoclimatic factor, K, for the average worst month for the region of operation. Appendix 1 of Ref. 2 gives the procedure for determining the geoclimatic factor based on fade measurements. In the absence of such measurements, the geoclimatic factor can be estimated using the ITU refractivity gradient data for the lowest 65 m of the atmosphere given in Ref. 1. The ITU refractivity data are available for 1%, 10%, 50%, 90%, and 99%, where the percentage is the percent of time that the value is not exceeded. The 1% map, shown in Figure 6.1, is used for determining K. Note that the data are concentrated in coastal areas as expected.

For reasonably accurate estimates of the probability of a multipath fade as a function of fade depth, the following procedure is used. First, estimate the geoclimatic factor from

$$K = 10^{-4.2-0.0029dN_1} \tag{6.12}$$

where dN_1 is read from Figure 6.3. The second step is to determine the path inclination in milliradians using

$$|\varepsilon_p| = \frac{|h_r - h_e|}{d} \qquad \text{mrad} \tag{6.13}$$

where h_e and h_r are the antenna heights above sea level in meters and d is the distance between the antennas in kilometers. The probability (in percent) of a fade of depth, A (dB), is then given by

$$p = Kd^{3.0}(1+|\varepsilon_p|)^{-1.2}10^{(0.033f-0.001h_L-A/10)} \quad \% \tag{6.14}$$

where h_L is the minimum of h_r and h_e, f is the frequency in GHz, and K is from the prior equation.

For a somewhat more precise estimate, the characteristics of the local terrain are taken into account. For this method, the value of K is again computed using data from Figure 6.3, but a different equation is used:

$$K = 10^{(-3.9-0.003dN_1)}s_a^{-0.42} \tag{6.15}$$

where s_a is the standard deviation of the terrain heights in meters within a 110- $\times$ 110-km area with 30-s (arc-seconds, or ~1-km) resolution. The ITU [2, p. 8] gives details on finding these data and how to apply it. The computation of the path inclination is the same as before:

$$|\varepsilon_p| = \frac{|h_r - h_e|}{d}$$

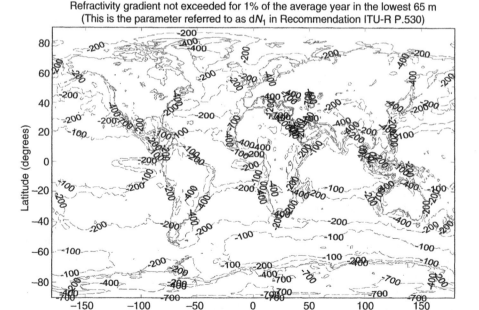

Refractivity gradient not exceeded for 1% of the average year in the lowest 65 m
(This is the parameter referred to as dN_1 in Recommendation ITU-R P.530)

Figure 6.3 Refractivity gradient data for estimating the geoclimatic factor. (Figure 12 from Ref. 1, courtesy of the ITU.)

The third and final step is again slightly different from the first method. The expression for the probability of exceeding a fade depth of A (dB) is

$$p = Kd^{3.2}(1+|\varepsilon_p|)^{-0.97} 10^{(0.32f-0.00085h_L-A/10)} \qquad \% \qquad (6.16)$$

These models for atmospheric multipath fade probabilities are valid up to 45 GHz and down to

$$f_{min} = 15/d \qquad \text{GHz}$$

where d is the link distance in km.

Example 6.2. What is the probability of a 4-dB or greater fade on a 15-km, horizontal link operating at 35 GHz on the U.S. gulf coast? Assume that the antennas are mounted on 20-m towers.

First the value of K must be determined:

$$K = 10^{-4.2-0.0029dN_1}$$

From Figure 6.3 the refractivity gradient (dN_1) is -200, so

$$K = 239.9 \times 10^{-6}$$

Since the path is horizontal, $\varepsilon_p = 0$ mrad. The expression for the probability of a fade is

$$p = Kd^{3.0}(1+|\varepsilon_p|)^{-1.2} 10^{(0.033f-0.001h_L-A/10)} \qquad \%$$

The probability of a 4-dB fade is then readily computed:

$$P_{4dB} = 4.4\%$$

So if a 4-dB fade margin were used, the link would be expected to drop out about 4.4% of the time. It is important to note that the multipath requires a calm, clear atmosphere and will not occur during rain. Thus a single fade margin for rain and/or multipath should be used. ☐

The remedies for atmospheric multipath include maintaining sufficient margins on the link or using antenna diversity to mitigate the fades. Using margin can be very costly because 30 dB or more may be required. Antenna diversity using two antennas is the preferred method for fixed links. For mobile links, either margin or tolerating a certain amount of outage time is usually used. It is worth noting that atmospheric multipath is not observed during moderate to heavy precipitation, so the link designer need only consider the larger of the two fade margins (rain or multipath), not both. Careful selection of the link path can also help mitigate multipath. Not straddling or parallel-ing bodies of water, running shorter links, and increasing the inclination angle all help to reduce the frequency and severity of multipath fades.

6.3 ATMOSPHERIC ATTENUATION

The atmosphere is comprised of gases, which absorb electromagnetic energy at various frequencies. Atmospheric attenuation due to gaseous absorption should not be confused with multipath or rain fades; it is a different mecha-nism. The gases of primary concern for microwave and millimeterwave systems are oxygen and water vapor. Similar to refraction, atmospheric losses also depend upon pressure, temperature, and water vapor content [12, 13]. For this reason, the effects can vary considerably with location, altitude, and the path slant angle. The atmosphere can be thought of as being comprised of hori-zontal layers at different altitudes, each having different water vapor and oxygen densities. Therefore, terrestrial links and earth-to-space/air links expe-rience different atmospheric effects and must be modeled differently [12].

Modeling the attenuation of RF signals by atmospheric gases is a well-established process that is outlined in Refs. 1 and 12. The recommended ap-proach is to use local atmospheric measurements along with the ITU models to predict the expected amount of absorption. Rather than using measured atmospheric data for a given location, the standard atmosphere parameters given in Ref. 1 are used throughout this chapter and indeed in most practical applications. For the frequencies in the range of 2–40 GHz, oxygen and water vapor are the dominant attenuation factors in the atmosphere.

For terrestrial links where both terminals are at or near the same altitude, the atmosphere can be treated as constant over the path. In this case it makes sense to characterize the absorption as a specific attenuation value in dB/km, which can be applied to the path distance to determine the total attenuation. The expression for the total loss due to atmospheric absorption on a terres-trial path is [2, equation (1)]

$$A = \gamma_a d \quad \text{dB} \tag{6.17}$$

where d is the line-of-sight distance between the terminals in km and γ_a is the specific attenuation of the atmosphere in dB/km. The specific attenuation of the atmosphere is given by the sum of the specific attenuation due to water vapor and that due to oxygen [12]:

$$\gamma_a = \gamma_o + \gamma_w \tag{6.18}$$

While the ITU gives formulas for these parameters [12], they are very long and tedious to evaluate. Instead of calculating the specific attenuation, it is common practice to use values extracted from a plot like Figure 6.4. The plot

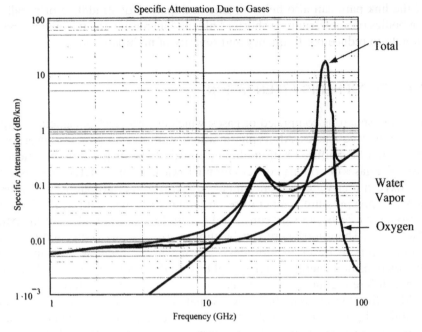

Figure 6.4 Specific attenuation due to atmospheric gases for a standard atmosphere.

in Figure 6.4 is for a horizontal path in a standard atmosphere at sea level. Figure 6.5 is the same plot over a much greater frequency range for both a standard reference atmosphere and for dry air, while Figure 6.6 provides the total attenuation through the entire atmosphere on a vertical path.

From the plots of γ, it can be seen that water vapor has an absorption line at 22 GHz. It can also be seen that dry air has an oxygen absorption line at 60 GHz. The areas between absorption lines are called *windows*. Note that at 1 GHz, $\gamma \sim 0.05$ dB/km, so even a 10-km path only incurs 0.5 dB of loss. The total loss due to atmospheric absorption for terrestrial paths is found by multiplying the path length by the specific attenuation. It is important to remember to multiply by 2 for radar applications since transmission must occur in two directions to complete the "link."

For slant paths, the calculations must allow for the variation in absorption with altitude. This is done by treating the atmosphere as a series of horizontal layers each with different pressure and temperature. While the actual absorption will be a fairly smooth function of altitude, treating it as discrete layers simplifies computation and provides a good estimate of the resulting attenuation. The ITU [12] provides detailed information on how to perform the calculations. A generalized approach for slant paths is given in Chapter 11, which is sufficient for most applications.

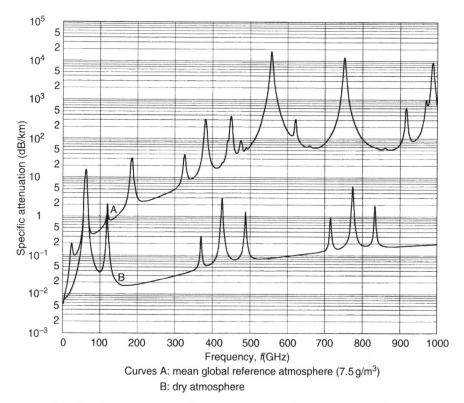

Figure 6.5 Specific attenuation of a standard atmosphere and of dry air versus frequency for a horizontal path. (Figure 1 from Ref. 12, courtesy of the ITU.)

For short-range communications, it is possible to operate at or near one of the absorption lines to improve frequency re-use or stealth since it reduces the distance a signal travels and therefore makes interference or stand-off signal detection unlikely.

Example 6.3. Consider a point-to-point mmw link that requires SNR = 10 dB and SIR = 25 dB for proper operation. What is the maximum operating distance and minimum frequency re-use distance of this system? Given:

$f = 60\,\text{GHz}, \lambda = 0.005\,\text{m}$

$G_{ANT} = 30\,\text{dB}$

$B = 50\,\text{MHz}$

$F = 5\,\text{dB}$

$P_T = 20\,\text{dBm}$

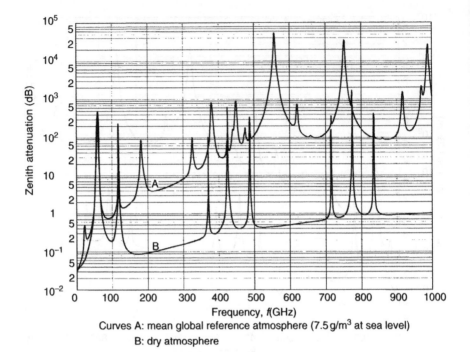

Curves A: mean global reference atmosphere (7.5 g/m³ at sea level)
B: dry atmosphere

Figure 6.6 Total atmospheric attenuation at zenith versus frequency for a standard reference atmosphere and for dry air. (Figure 3 from Ref. 12, courtesy of the ITU.)

The noise floor is kT_0BF, so

$$N = -174\,\text{dB-W/Hz} + 77\,\text{dB-Hz} + 5\,\text{dB} = -92\,\text{dBm}$$

Thus

$$S_{min} = -92 + 10 = -82\,\text{dBm}$$

So we have

$$P_T - S_{min} = 20\,\text{dBm} - (-82\,\text{dBm}) = 102\,\text{dB}$$

of margin for path loss. The total path loss is

$$PL = -20\log(G\lambda/(4\pi d)) + 15\,\text{dB/km} \cdot d$$

The maximum interference signal is $S_{min} - 25\,\text{dB} = -107\,\text{dBm}$. Since $P_T = 20\,\text{dBm}$, 127 dB of path loss is required to mitigate interference. From the solid

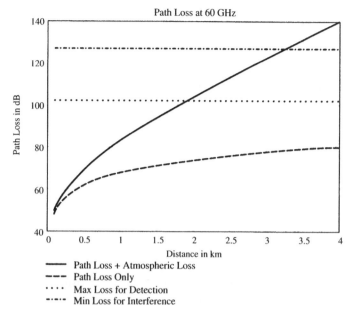

Figure 6.7   Path loss for a 60-GHz communication link showing the impact of atmospheric absorption.

curve in Figure 6.7, it can be estimated that the maximum operating distance is 1.8 km and the minimum separation to avoid interference will be about 3.3 km (assuming main beam intersection of the transmitter and the victim receiver).

Figure 6.7 shows the path loss at 60 GHz for this system both with and without the atmospheric absorption. It is clear that the atmospheric absorption plays a significant role. It increases the required EIRP for communication (or reduces the communication distance), but it also permits frequency reuse as close as 3.3 km. ☐

6.4   LOSS FROM MOISTURE AND PRECIPITATION

Moisture in the atmosphere takes many forms, some of which play a significant role in attenuating electromagnetic waves. The effects are, of course, frequency-dependent. Precipitation in the form of rain, freezing rain, or wet snow is a significant problem above about 10 GHz and is treated separately in Chapter 10. Water vapor is essentially a gas and was treated in Section 6.3. Suspended droplets of moisture, such as clouds or fog, is neither precipitation nor water vapor and is therefore treated separately. Dry snow and dust also presents a unique challenge to the modeler.

6.4.1 Fog and Clouds

The ITU provides a recommended model for fog or cloud attenuation [14] that is valid up to 200 GHz. Application of this model and a few representative design curves for planning are presented in this section. The specific attenuation due to clouds or a fog bank is

$$\gamma_c = K_l M$$

where

γ_c is the specific attenuation of the cloud in dB/km
K_l is the specific attenuation coefficient of the cloud in (dB/km)/(g/m³)
M is the liquid water density of the cloud in g/m³

The ITU gives the liquid water density of a cloud or fog as

$$M = 0.05 \text{ g/m}^3 \text{ for medium fog (300-m visibility)}$$
$$M = 0.5 \text{ g/m}^3 \text{ for dense fog (50-m visibility)}$$

The mathematical model for the specific attenuation coefficient is given as

$$K_l = \frac{0.819 f}{\varepsilon''(1+\eta^2)} \quad (\text{dB/km})/(\text{g/m}^3) \qquad (6.19)$$

where f is the frequency in GHz and

$$\eta = \frac{2+\varepsilon'}{\varepsilon''}$$

$$\varepsilon''(f) = \frac{f(\varepsilon_0 - \varepsilon_1)}{f_p\left[1+\left(\dfrac{f}{f_p}\right)^2\right]} + \frac{f(\varepsilon_1 - \varepsilon_2)}{f_s\left[1+\left(\dfrac{f}{f_s}\right)^2\right]}$$

$$\varepsilon'(f) = \frac{(\varepsilon_0 - \varepsilon_1)}{\left[1+\left(\dfrac{f}{f_p}\right)^2\right]} + \frac{(\varepsilon_1 - \varepsilon_2)}{\left[1+\left(\dfrac{f}{f_s}\right)^2\right]} + \varepsilon_2$$

$$\varepsilon_0 = 77.6 + 103.3(\theta - 1)$$

$$\varepsilon_1 = 5.48$$

$$\varepsilon_2 = 3.51$$

$$\theta = \frac{300}{T}$$

Here the temperature, T, is in kelvins. The principal and secondary relaxation frequencies are given by

$$f_p = 20.09 - 142(\theta - 1) + 294(\theta - 1)^2 \qquad \text{GHz}$$

$$f_s = 590 - 1500(\theta - 1) \qquad \text{GHz}$$

Figure 6.8 shows a plot of the specific attenuation coefficient, K_l, versus frequency for –10°C, 0°C, and 20°C. For clouds, the 0°C curve should be used. Values from this curve can be used along with liquid water density to estimate the specific attenuation (dB/km) due to clouds or fog. Figure 6.9 is a plot of the specific attenuation versus frequency of clouds or fog for a liquid water density of 0.5 g/m³. In the absence of specific measured liquid water density, these curves can be used to estimate the specific attenuation of clouds or fog. Here again, the 0°C curve should be used for clouds.

To determine the total attenuation through the atmosphere due to clouds, the elevation angle and the total columnar content of liquid water, L, must be known. The total columnar content of liquid water may be estimated for a given probability using the plots provided in Ref. 14. The worst-case value is 1.6 kg/m². The expression for the total cloud attenuation is

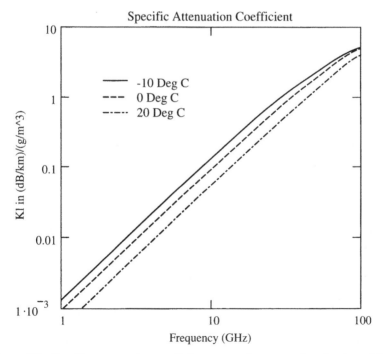

Figure 6.8 Specific attenuation coefficient versus frequency for clouds and fog.

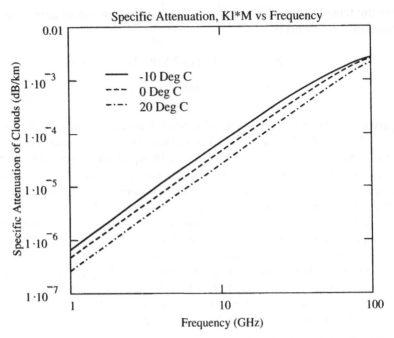

Figure 6.9 Specific attenuation versus frequency for clouds and fog, liquid water density of 0.5 g/m³.

$$A = \frac{LK_l}{\sin(\theta)} \qquad dB \qquad for \ 90° \geq \theta \geq 5° \qquad (6.20)$$

where

L is the total columnar liquid water in kg/m²
K_l is the specific attenuation coefficient in dB/km
θ is the elevation angle of the path

Note that the units of L and K_l are such that the end result will be in dB. Figure 6.10 shows a plot of the worst-case cloud attenuation through the atmosphere for several different elevation angles. It can be seen that cloud attenuation starts to become a consideration above about 10 GHz.

It is intuitive that fog will not be present during a heavy rain. Therefore if the attenuation due to fog is less than the required rain margin, only the rain margin need be included in the link budget. While clouds are likely to be present during rain, the rain models are based on empirical measurements, which include any cloud attenuation, so again the cloud attenuation need not be included in the link budget if it is smaller than the rain margin (or vice versa).

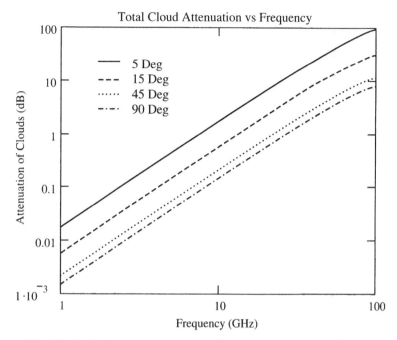

Figure 6.10 Total cloud attenuation versus frequency for several elevation angles using worst-case cloud density.

Example 6.4. Consider a ground to air communication link operating at 20 GHz. If the aircraft is a high-altitude craft, operating well above the clouds, what is the worst-case additional path loss due to cloud cover? Assume that the minimum operational elevation angle is 15 degrees.

Using the expression for cloud attenuation,

$$A = \frac{LK_l}{\sin(\theta)} \quad \text{dB} \quad \text{for } 90° \geq \theta \geq 5°$$

K_1 at 20 GHz and 0°C can be computed as

$$K_1 = 0.36 \text{ m}^2/\text{kg}$$

The worst-case value for L was given as

$$L = 1.6 \text{ kg}/\text{m}^2$$

and

$$\theta = 15°$$

The worst-case attenuation due to heavy cloud cover can then be predicted to be

$$A = 2.23 \, dB$$

which is consistent with the plot in Figure 6.10. ☐

Example 6.5. Consider a 10-GHz ship-to-shore communication link. If the maximum communication distance is 12 km, what is the worst-case loss due to fog?
Using

$$M = 0.5 \, g/m^3$$

for the liquid water density and

$$T = -10°C$$

the specific attenuation coefficient is computed as

$$K_1 = 0.132 \, m^2/kg$$

and the resulting specific attenuation is

$$\gamma = 0.066 \, dB/km$$

Multiplying the specific attenuation by the maximum communication distance of 12 km results in a maximum attenuation of

$$A = 0.79 \, dB$$

which is hardly enough loss to be concerned about, especially when compared to the likely rain loss. Depending upon the location, fog may occur much more often than rain, so each case must be treated separately. ☐

6.4.2 Snow and Dust

The attenuation of electromagnetic waves due to snow or dust [13] is predominantly a function of the moisture content of the particles. For that reason, the attenuation effects are expected to be less than those due to rain. Specific procedures for modeling the effects of snow or dust are few and far between and since, to the author's knowledge, none have gained wide acceptance, a model is not provided in this book. The recommendation is that the rain attenuation models more than cover the snow and dust cases and since the effects are mutually exclusive, planning for dry snow or dust is superfluous in most cases. Note that snow accumulation on antennas and radomes is a separate phenomenon from the effect of atmospheric snow.

The effect of hail will depend upon its moisture content and the size of the hailstones relative to the wavelength of the wave. Here again, definitive models are difficult to come by. The relative infrequency of hailstorms and their association with thunderstorms makes modeling the effects of hail independently from rain unnecessary for most applications.

6.5 SUMMARY

There is a wide diversity of atmospheric effects that must be considered based on the frequency of operation, the path elevation, and the local environment. Atmospheric refraction occurs when the density gradient of the atmosphere moves the apparent radio horizon beyond the geometric horizon for nearly horizontal links. The effect is most often compensated for by using a 4/3 earth radius model for determining the distance to the horizon. If detailed atmospheric data are available, it is possible to find a more accurate representation of the radio horizon. A related phenomenon is ducting, which is attributed to water vapor in the atmosphere. A duct may either exist at the surface or be elevated as much as 1500 m. The effect of a duct is to trap the RF wave and prevent the usual geometric spreading loss, resulting in excessively strong receive signals if the receiver is located in the same duct with the transmitter. If the receiver is not in the same duct, the result will be a significant fade, which can be treated as a multipath fade.

Atmospheric multipath models are based on empirical measurements (and hence they include cross-duct fades). The current ITU model is similar to previously used models and has been shown to match empirical data very well [10]. It consists of two different models for low-probability fades and a third, more detailed model that is validated for all fade probabilities.

The third major effect of the atmosphere is absorption and attenuation of RF waves. The dominant absorption effects of the atmosphere are due to oxygen and water vapor. The absorption is generally described by a specific attenuation in dB/km for terrestrial paths and by a total attenuation as a function of elevation angle for slant paths that exit the troposphere.

There are three forms of water that are of interest to the propagation analyst: precipitation, water vapor, and suspended water droplets, forming clouds or fog. Each has unique properties and affects electromagnetic waves differently. The effect of clouds and fog is rarely significant, but it can be a factor, particularly for log terrestrial paths or for satellite links. The ITU models for determining the specific attenuation due to fog and for determining the total attenuation due to clouds on a slant path were both described. The attenuation of RF waves by snow and dust is directly related to the moisture content of the particles and is small relative to rain attenuation.

REFERENCES

1. ITU-R Recommendations, *The radio refractive index: Its formula and refractivity data*, ITU-R P.453-8, Geneva, 2001.
2. ITU-R Recommendations, *Propagation data and prediction methods required for the design of terrestrial line-of-sight system*, ITU-R P.530-9, Geneva, 2001.
3. J. D. Parsons, *The Mobile Radio Propagation Channel*, 2nd ed., Wiley, West Sussex, 2000, p. 29.
4. ITU-R Recommendations, *Reference standard atmospheres*, ITU-R P.835-3, Geneva, 1999.
5. J. D. Parsons, *The Mobile Radio Propagation Channel*, 2nd ed., Wiley, West Sussex, 2000, p. 28.
6. J. D. Parsons, *The Mobile Radio Propagation Channel*, 2nd ed., Wiley, West Sussex, 2000, p. 31.
7. W. T. Barnett, Multipath propagation at 4, 6, and 11 GHz, *The Bell System Technical Journal*, Vol. 51, No. 2, February 1972, pp. 311–361.
8. K. Morita, Prediction of Rayleigh fading occurrence probability of line-of-sight microwave links, *Review of Electrical Communication Laboratories*, Vol. 18, December 1970, pp. 310–321.
9. R. L. Olsen and B. Segal, New techniques for predicting the multipath fading distribution on VHF/UHF/SHF terrestrial line-of-sight links in Canada, *Canadian Journal of Electrical and Computing Engineering*, Vol. 17, No. 1, 1992, pp. 11–23.
10. R. L. Olsen, T. Tjelta, L. Martin, and B. Segal, Worldwide techniques for predicting the multipath fading distribution on terrestrial L.O.S. links: Comparison with regional techniques, *IEEE Transactions on Antennas and Propagation*, Vol. 51, No. 1, January 2003, pp. 23–30.
11. ITU-R Recommendations, *Propagation data and prediction methods required for the design of terrestrial line-of-sight systems*, ITU-R P.530-8, Geneva, 1999.
12. ITU-R Recommendations, *Attenuation by atmospheric gases*, ITU-R P.676-5, Geneva, 2001.
13. S. A. A. Abdulla, H. M. Al-Rizzo, and M. M. Cyril, Particle size distribution of Iraqi sand and dust storms and their influence on microwave communication systems, *IEEE Transactions on Antennas and Propagation*, Vol. 36, No. 1, January 1988, pp. 114–126.
14. ITU-R Recommendations, *Attenuation due to clouds and fog*, ITU-R P.840-3, Geneva, 1999.

EXERCISES

1. A point-to-point communication system has one antenna mounted on a 5-m tower and antenna at the other end of the link is mounted on a 10-m tower.

   (a) What is the distance to the horizon for each antenna?

   (b) What is the maximum possible link distance if only line-of-sight is considered?

(c) For the maximum possible link distance found in part b, what is the expected amount of water vapor and oxygen loss?

(d) How would these results change if the path were located in a semi-tropical climate and over a large body of water?

2. What is the probability of a 10-dB or greater multipath fade for a point-to-point communications link that parallels the coast in the Pacific Northwest region of the United States? The link is operating at 28 GHz over a 10-km distance, and the antennas are mounted on 30-m towers.

3. Repeat problem 2 if the link is located along the California coast.

4. Consider a 38-GHz, point-to-point communication system operating on the east coast of Florida. If the antennas are mounted on 10-m towers and the communication distance is 8 km, how much multipath fade margin is required if a 98% multipath availability is required?

5. Consider a ground-to-air communication link operating at 18 GHz. If the nominal elevation angle to the aircraft is 30 degrees, what is the worst-case cloud attenuation that can be expected?

6. Approximately how much attenuation will be experienced by a 0.8-km, 1.9-GHz PCS link operating near the Florida coast:

(a) Due to atmospheric absorption?

(b) Due to fog?

CHAPTER 7

Near-Earth Propagation Models

7.1 INTRODUCTION

Many applications require RF or microwave propagation from point to point very near the earth's surface and in the presence of various impairments. Examples of such applications include cellular telephones, public service radio, pagers, broadcast television and radio stations, and differential GPS transmitters. Propagation loss over terrain, foliage, and/or buildings may be attributed to various phenomena, including diffraction, reflection, absorption, or scattering. In this chapter, several different models are considered for determining the median (50%) path loss as a function of distance and conditions. These models are all based on measurements (sometimes with theoretical extensions) and represent a statistical mean or median of the expected path loss. In the next chapter, the effects of multipath and shadowing are examined in detail. Much of the data collection for near-earth propagation impairment has been done in support of mobile VHF communications and, more recently, mobile telephony (which operates between 800 MHz and 2 GHz). Thus many of the models are focused in this frequency range. While for the most part, models based on this data are only validated up to 2 GHz, in practice they can sometimes be extended beyond that if required. Some of the recent measurement campaigns and models have specifically targeted higher-frequency operation, particularly the updated ITU models. In their tutorial paper, Bertoni et al. [1] provide an excellent overview of the subject of near-earth propagation modeling.

7.2 FOLIAGE MODELS

Most terrestrial communications systems require signals to pass over or through foliage at some point. This section presents a few of the better-known foliage models. These models provide an estimate of the additional attenuation due to foliage that is within the line-of-sight (LOS) path. There are of

Introduction to RF Propagation, by John S. Seybold
Copyright © 2005 by John Wiley & Sons, Inc.

134

course, a variety of different models and a wide variation in foliage types. For that reason, it is valuable to verify a particular model's applicability to a given region based on historical use or comparison of the model predictions to measured results.

7.2.1 Weissberger's Model

Weissberger's modified exponential decay model [2, 3] is given by

$$L(dB) = \begin{cases} 1.33 F^{0.284} d_f^{0.588}, & 14 < d_f \leq 400 \text{ m} \\ 0.45 F^{0.284} d_f, & 0 < d_f \leq 14 \text{ m} \end{cases} \qquad (7.1)$$

where

d_f is the depth of foliage along the LOS path in meters
F is the frequency in **GHz**

The attenuation predicted by Weissberger's model is in addition to free-space (and any other nonfoliage) loss. Weissberger's modified exponential decay model applies when the propagation path is blocked by dense, dry, leafed trees. It is important that the foliage depth be expressed in meters and that the frequency is in GHz. Blaunstein [3] indicates that the model covers the frequency range from 230 MHz to 95 GHz.

7.2.2 Early ITU Vegetation Model

The early ITU foliage model [4] was adopted by the CCIR (the ITU's predecessor) in 1986. While the model has been superseded by a more recent ITU recommendation, it is an easily applied model that provides results that are fairly consistent with the Weissberger model. The model is given by

$$L(dB) = 0.2 F^{0.3} d_f^{0.6} \qquad dB \qquad (7.2)$$

where

F is the frequency in **MHz**
d_f is the depth of the foliage along the LOS path in meters

Figures 7.1 and 7.2 show comparisons of the Weissberger and ITU models, for foliage depths of 5, 20, 50, and 100 m. Note that the frequency scale is in GHz for each plot, but the frequency used in the model is MHz for the ITU model as specified. The plots indicate a moderate variation between the models, particularly as frequency increases. The amount of foliage loss is monotonically increasing with foliage depth and frequency as expected.

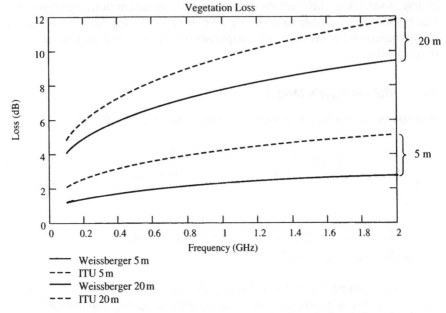

Figure 7.1 Vegetation loss versus frequency for 5- and 20-m foliage depth.

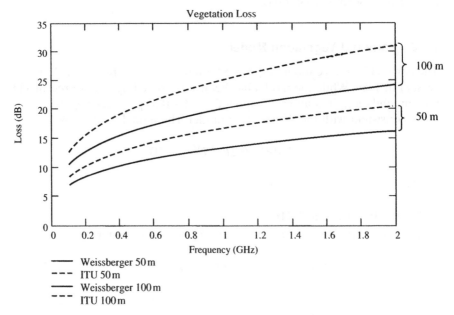

Figure 7.2 Vegetation loss versus frequency for 50- and 100-m foliage depth.

Example 7.1. Consider a system with the following parameters:

$$d = 1 \text{ km}, \qquad f = 1 \text{ GHz}$$

with 12 m of trees in the LOS and vegetation (leaves) present. What is the total predicted median path loss for this system (excluding antenna gains).

First, compute the free-space loss (FSL):

$$\text{FSL} = -20 \log \left(\frac{\lambda}{4 \pi d} \right) = 92.44 \text{ dB}$$

since

$$\lambda = 0.3 \text{ m}, \qquad d = 1000 \text{ m}$$

For the Weissberger model, $d_f < 14 \text{ m}$, so

$$L_{dB} = 0.45 (1 \text{ GHz})^{0.284} d_f$$

With $d_f = 12 \text{ m}$, this yields

$$L_{dB} = 5.4 \text{ dB}$$

Thus the total median path loss predicted by the Weissberger model is

$$L_{50} = 97.8 \text{ dB}$$

For the ITU model, the loss is given by

$$L_{dB} = 0.2 (1000 \text{ MHz})^{0.3} d_f^{0.6}$$

With $d_f = 12 \text{ m}$, the loss due to foliage is found to be

$$L_{dB} = 7.06 \text{ dB}$$

So the total median path loss predicted by the ITU model is

$$L_{50} = 99.5 \text{ dB} \quad \square$$

7.2.3 Updated ITU Vegetation Model

The current ITU models are fairly specific and do not cover all possible scenarios. Nonetheless they are valuable and represent a recent consensus. One of the key elements of the updated model, which should also be considered in applying other models is that there is a limit to the magnitude of the attenuation due to foliage, since there will always be a diffraction path over and/or around the vegetation [5].

7.2.3.1 Terrestrial Path with One Terminal in Woodland

The scenario covered by this model is shown in Figure 7.3. The model for the excess attenuation due to vegetation is

$$A_{ev} = A_m \left[1 - e^{d\gamma/A_m} \right] \quad \text{dB} \tag{7.3}$$

where

  d is the length of the path that is within the woodland in *meters*
  γ is the specific attenuation for very short vegetative paths (dB/m)
  A_m is the maximum attenuation for one terminal within a specific type and depth of vegetation (dB)

The excess attenuation due to vegetation is, of course, added to the free-space loss and the losses from all other phenomena to determine the total predicted path loss. Some typical values for the specific attenuation are plotted versus frequency in Figure 7.4.

7.2.3.2 Single Vegetative Obstruction

If neither end of the link is within woodland, but there is vegetation within the path, the attenuation can be modeled using the specific attenuation of the vegetation. For this model to apply, the vegetation must be of a single type, such as a tree canopy, as opposed to a variety of vegetation. When the frequency is at or below 3 GHz, the vegetation loss model is

$$A_{et} = d\gamma \tag{7.4}$$

where

  d is the length of the path that is within the vegetation (in meters)
  γ is the specific attenuation for short vegetative paths (dB/m)
  $A_{et} \leq$ the lowest excess attenuation for any other path (dB)

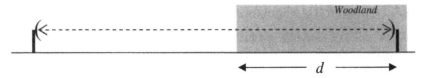

Figure 7.3 Propagation path with one terminal in woodland for ITU model.

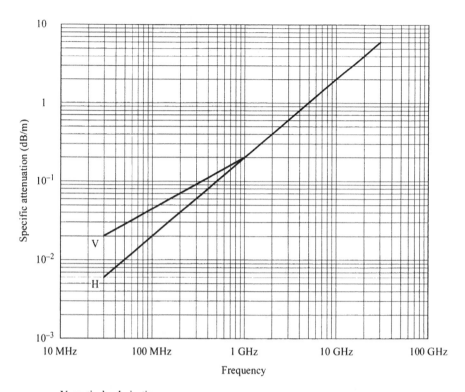

V: vertical polarization
H: horizontal polarization

Figure 7.4 Specific attenuation due to vegetation versus frequency. (Figure 2 from Ref. 5, courtesy of the ITU.)

The restriction on A_{et} ensures that if the vegetation loss is very large, any alternate paths such as a diffraction path will determine the path loss. The ITU indicates that this model is an approximation and will tend to overestimate the actual foliage attenuation.

The updated ITU model does not provide for coverage between 3 and 5 GHz other than the *one terminal in woodland model*. Above 5 GHz, the updated ITU model is based on the type of foliage, the depth of the foliage, and the illuminated area of the foliage. The excess attenuation due to vegetation is given by

$$A_{veg} = R_{\infty}d + k\left[1 - e^{(-R_0 + R_{\infty})\frac{d}{k}}\right] \quad \text{dB} \tag{7.5}$$

where

$R_0 = af$, the initial slope
$R_{\infty} = b/f^c$, the final slope

TABLE 7.1   Parameters for Updated ITU Model

Parameter	In Leaf	Out of Leaf
a	0.2	0.16
b	1.27	2.59
c	0.63	0.85
k_0	6.57	12.6
R_f	0.0002	2.1
A_0	10	10

Source: Table 1 from Ref. 5, courtesy of the ITU.

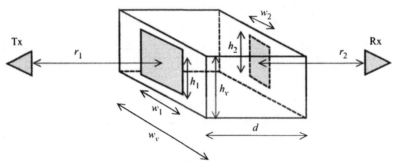

Figure 7.5   Geometry of minimum illuminated vegetation area. (Figure 3 from Ref. 5, courtesy of the ITU.)

f is the frequency of operation (in GHz)

a, b, and c are given in Table 7.1

and

$$k = k_0 - 10\log\left[A_0\left(1 - e^{-A_{min}/A_0}\right)\left(1 - e^{-R_{ff}}\right)\right] \qquad (7.6)$$

where k_0, A_0, and R_f are also given in Table 7.1. A_{min} is the illumination area, which is computed based on the size of the vegetation patch and the illumination pattern of the antenna. The definition of A_{min} is the smallest height by the smallest width of illuminated clutter. The height and width are determined by the height and width of the clutter patch and by the height and width of the transmit and the receive antenna patterns (3-dB beamwidth) where they intersect the vegetation. Figure 7.5 shows the geometry of h_1, h_2, h_v, w_1, w_2, and w_v:

$$A_{min} = \min(h_1, h_2, h_v) \times \min(w_1, w_2, w_v) \qquad (7.7)$$

The expression for A_{min} can also be written in terms of the distances to the vegetation and the elevation and azimuth beamwidths of the antennas.

$$A_{min} = \min\left(2r_1 \tan\left(\frac{\phi_T}{2}\right), 2r_2 \tan\left(\frac{\phi_R}{2}\right), h_v \right) \times \min\left(2r_1 \tan\left(\frac{\theta_T}{2}\right), 2r_2 \tan\left(\frac{\theta_R}{2}\right), h_v \right)$$

(7.8)

where

r_1 and r_2 are the distances to the vegetation as shown in Figure 7.5
ϕ_T and ϕ_R are the transmit and receive elevation beamwidths
θ_T and θ_R are the transmit and receive azimuth beamwidths
h_v and w_v are the height and width of the vegetation patch

7.3   TERRAIN MODELING

For ground-based communications, the local terrain features significantly affect the propagation of electromagnetic waves. Terrain is defined as the natural geographic features of the land over which the propagation is taking place. It does not include vegetation or man-made features. When the terrain is very flat, only potential multipath reflections and earth diffraction, if near the radio horizon, need to be considered. Varied terrain, on the other hand, can produce diffraction loss, shadowing, blockage, and diffuse multipath, even over moderate distances. The purpose of a terrain model is to provide a measure of the median path loss as a function of distance and terrain roughness. The variation about the median due to other effects are then treated separately.

7.3.1   Egli Model

While not a universal model, the Egli model's ease of implementation and agreement with empirical data make it a popular choice, particularly for a first analysis. The Egli model for median path loss over irregular terrain is [4, 6, 7]

$$L_{50} = G_b G_m \left[\frac{h_b h_m}{d^2}\right]^2 \beta$$

(7.9)

where

G_b is the gain of the base antenna
G_m is the gain of the mobile antenna
h_b is the height of the base antenna

h_m is the height of the mobile antenna

d is the propagation distance

$\beta = (40/f)^2$, where f is in MHz

Note that the Egli model provides the entire path loss, whereas the foliage models discussed earlier provided the loss in addition to free-space loss. Also note that the Egli model is for irregular terrain and does not address vegetation. While similar to the ground-bounce loss formula, the Egli model is not based on the same physics, but rather is an empirical match to measured data [4]. By assuming a log-normal distribution of terrain height, Egli generated a family of curves showing the terrain factor or adjustment to the median path loss for the desired fade probability [6]. This way the analyst can determine the mean or median signal level at a given percentage of locations on the circle of radius d. Stated another way, the Egli model provides the median path loss due to terrain loss. If a terrain loss point other than the median (50%) is desired, the adjustment factor in dB can be inferred from Figure 7.6.

Example 7.2. Determine the median terrain loss for a 1-km link operating at 100 MHz if the antenna heights are 20 m and 3 m, using the Egli model for terrain loss.

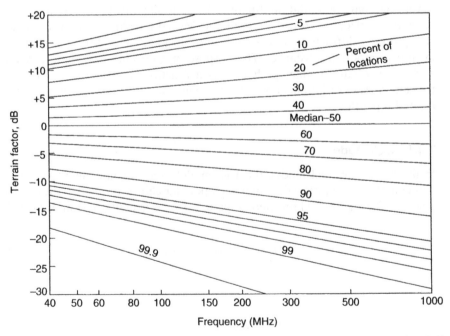

Figure 7.6 Terrain factor versus frequency for different probabilities for the Egli model. (Figure 3.20 from Ref. 4, courtesy of Wiley.)

Applying equation (7.9),

$$L_{50} = -10\log\left(\frac{20 \cdot 3}{10^6} \cdot (0.4)^2\right)$$

and thus

$$L_{50} = 112.4 \, \text{dB}$$

If the 90th percentile is desired (i.e., the level of terrain loss that will be exceeded 10% of the time), approximately 10 dB would be added to the L_{50} value according to Figure 7.6, so

$$L_{90} = 122.2 \, \text{dB} \quad \square$$

The Egli model provides a nice, closed-form way to model terrain effects, but since it is a one-size-fits-all model, it should not be expected to provide precise results in all situations. For detailed planning, there are software packages available that use DTED (Digitized Terrain Elevation Data) or similar terrain data and model the expected diffraction loss on a given path. Such models are ideal for planning fixed links, but are of limited utility for mobile links. One exception is the Longley-Rice model, which provides both point-to-point and area terrain loss predictions.

7.3.2 Longley–Rice Model

The Longley–Rice model is a very detailed model that was developed in the 1960s and has been refined over the years [8–10]. The model is based on data collected between 40 MHz and 100 GHz, at ranges from 1 to 2000 km, at antenna heights between 0.5 and 3000 m, and for both vertical and horizontal polarization. The model accounts for terrain, climate, and subsoil conditions and ground curvature. Blaunstein [8] provides a detailed description of the model, while Parsons [9] provides details determining the inputs to the model. Because of the level of detail in the model, it is generally applied in the form of a computer program that accepts the required parameters and computes the expected path loss. At the time of this writing, the U.S. National Telecommunications and Information Administration (NTIA) provides one such program on its website [11] free of charge. Many commercial simulation products include the Longley–Rice model for their terrain modeling. As indicated in the previous section, the Longley–Rice model has two modes, point-to-point and area. The point-to-point mode makes use of detailed terrain data or characteristics to predict the path loss, whereas the area mode uses general information about the terrain characteristics to predict the path loss.

7.3.3 ITU Model

The ITU terrain model is based on diffraction theory and provides a relatively quick means of determining a median path loss [12]. Figure 7.7 shows three plots of the expected diffraction loss due to terrain roughness versus the normalized terrain clearance. Curve B is the theoretical knife-edge diffraction curve. Curve D is the theoretical smooth-earth loss at 6.5 GHz using a 4/3 earth radius. The curve labeled A_d is the ITU terrain loss model over intermediate terrain. Each of these curves represents the excess terrain loss, beyond free-space loss. The ITU terrain loss model is given by

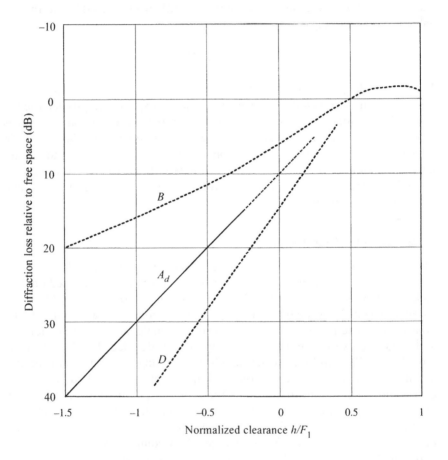

B: theoretical knife-edge loss curve
D: theoretical smooth spherical Earth loss curve, at 6.5 GHz and $k_e = 4/3$
A_d: empirical diffraction loss based on equation (2) for intermediate terrain
h: amount by which the radio path clears the Earth's surface
F_1: radius of the first Fresnel zone

Figure 7.7 Additional loss due to terrain diffraction versus the normalized clearance. (Figure 1 from Ref. 12, courtesy of the ITU.)

$$A_d = -20h/F_1 + 10 \quad \text{dB} \tag{7.10}$$

where h is the height difference between the most significant path blockage and the line-of-sight path between the transmitter and the receiver. If the blockage is above the line of sight, then h is negative. F_1 is the radius of the first Fresnel zone (Fresnel zones are discussed in Chapter 8) and is given by

$$F_1 = 17.3\sqrt{\frac{d_1 d_2}{fd}} \quad \text{m} \tag{7.11}$$

where

d_1 and d_2 are the distances from each terminal to the blockage in kilometers

d is the distance between the terminals in km

f is the frequency in GHz

The ratio h/F_1 is the normalized terrain clearance ($h/F_1 < 0$ when the terrain blocks the line of sight). This model is generally considered valid for losses above 15 dB, but it is acceptable to extrapolate it to as little as 6 dB of loss as shown in Figure 7.7. The other two curves shown represent extremes of clear terrain and very rough terrain, so they provide insight into the variability that can be expected for any given value of normalized clearance.

Example 7.3. A VHF military vehicle communication system needs to communicate with other military vehicles over a distance of 3 km over fairly rough terrain (±2 m) at 100 MHz. The antenna is a 1.5-m whip mounted on the vehicle, approximately 2 m above the ground. How much terrain loss should be expected over and above the free-space loss?

The center of radiation for the antenna is 2.75 m above the ground. If a flat-earth model is used, then the maximum terrain height of +2 m results in a minimum antenna height above the terrain of

$$h = 0.75 \text{ m}$$

In the absence of specific information about the location of any blockage within the line of sight, assume that the blockage occurs at the midpoint of the path, $d/2$. The expression for F_1 reduces to

$$F_1 = 17.3\sqrt{\frac{d}{4f}} \quad \text{m}$$

Next, by substituting $d = 3$ and $f = 0.1$ the value of F_1 is found to be

$$F_1 = 47.4$$

The normalized terrain clearance is $h/F_1 = 0.0158$ and the terrain attenuation is

$$A_d = 9.7 \text{ dB}$$

which is consistent with Figure 7.7. The plot also shows that terrain loss in the region of 6–15 dB might be reasonably expected. □

7.4 PROPAGATION IN BUILT-UP AREAS

Propagation of electromagnetic waves through developed areas from suburban to dense urban is of considerable interest, particularly for mobile telephony. This is a vast subject with numerous papers and models available. The actual propagation of RF though an urban environment is dependent upon frequency, polarization, building geometry, material structure, orientation, height, and density. This section treats propagation between elevated base stations and mobiles that are at street level in urban and suburban areas [13]. The goal is to determine the median path loss or RSL as a function of the distance, d, so that the required multipath fading models can then be applied (Chapter 8). The median value depends heavily upon the size and density of the buildings, so classification of urban terrain is important. The models discussed are the Young, Okumura, Hata, and Lee models.

7.4.1 Young Model

The Young data were taken in New York City in 1952 and covers frequencies of 150–3700 MHz [14, 15]. The curve presented in Figure 7.8 displays an inverse fourth-power law behavior, similar to the Egli model. The model for Young's data is

$$L_{50} = G_b G_m \left(\frac{h_b h_m}{d^2} \right)^2 \beta \tag{7.12}$$

where β is called the *clutter factor* and is not the same β used in the Egli model! This β is also distinct from the β sometimes used for building volume over a sample area in classification [16]. From Young's measurements, β is approximately 25 dB for New York City at 150 MHz. The data in Figure 7.8 suggests that a log-normal fit to the variation in mean signal level is reasonable.

7.4.2 Okumura Model

The Okumura model is based on measurements made in Tokyo in 1960, between 200 and 1920 MHz [17–20]. While not representative of modern U.S. cities, the data and model are still widely used as a basis of comparison. The

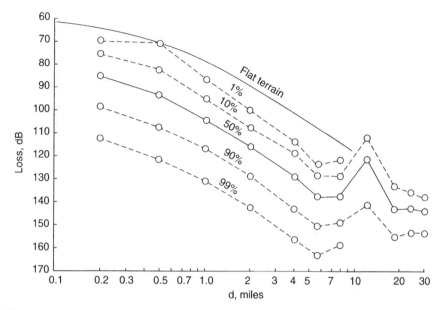

Figure 7.8   Results of Young's measurement of path loss versus distance in miles in Manhattan and the Bronx at 150 MHz. (Figure 7.1 from Ref. 15, courtesy of Artech House.)

model is empirical, being based solely on the measured data. The actual path loss predictions are made based on graphs of Okumura's results, with various correction factors applied for some parameters.

For the Okumura model, the prediction area is divided into terrain categories: *open area*, *suburban area*, and *urban area*. The open-area model represents locations with open space, no tall trees or buildings in the path, and the land cleared for 300–400 m ahead (i.e., farmland). The suburban area model represents a village or a highway scattered with trees and houses, some obstacles near the mobile, but not very congested. The urban area model represents a built-up city or large town with large buildings and houses with two or more stories, or larger villages with close houses and tall thickly grown trees. The Okumura model uses the urban area as a baseline and then applies correction factors for conversion to other classifications. A series of terrain types is also defined. Quasi-smooth terrain is the reference terrain and correction factors are applied for other types of terrain. Okumura's expression for the median path loss is

$$L_{50}(\text{dB}) = L_{FSL} + A_{mu} - H_{tu} - H_{ru} \qquad (7.13)$$

where

L_{FSL} is the free-space loss for the given distance and frequency

A_{mu} is the median attenuation relative to free-space loss in an urban area, with quasi-smooth terrain, base station effective height $h_{te} = 200\,\text{m}$, and mobile antenna height $h_{re} = 3\,\text{m}$; the value of A_{mu} is a function of both frequency and distance

H_{tu} is the base station height gain factor

H_{ru} is the mobile antenna height gain factor

The signs on the gain factors are very important. Some works have reversed the signs on the H terms, which will of course lead to erroneous results. If in doubt, check the results using known test cases, or engineering judgment. For instance, if increasing the antenna height increases the median path loss, then the sign of the antenna height correction factor is clearly reversed.

Figure 7.9 shows plots of A_{mu} versus frequency for various distances. Figure 7.10 shows the base station height gain factor in urban areas versus effective

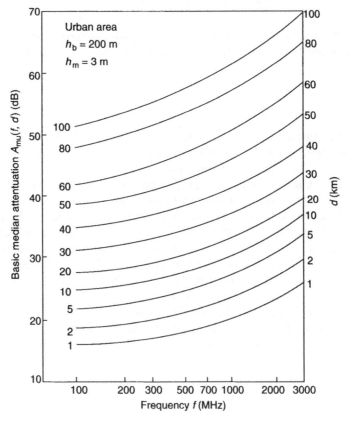

Figure 7.9 Plot of A_{mu} versus frequency for use with the Okumura model. (Figure 4.7 Ref. 13, courtesy of Wiley.)

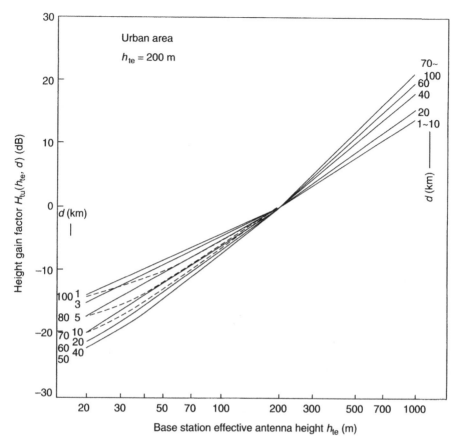

Figure 7.10 Plot of H_{tu}, the base station height correction factor, for the Okumura model. (Figure 4.8 from Ref. 13, courtesy of Wiley.)

height for various distances, while Figure 7.11 shows the vehicle antenna height gain factor versus effective antenna height for various frequencies and levels of urbanization. Figure 7.12 shows how the base station antenna height is measured relative to the mean terrain height between 3 and 15 km in the direction of the receiver.

Example 7.4. Consider a system with the following parameters:

$$h_t = 68 \text{ m}$$

$$h_r = 3 \text{ m}$$

$$f = 870 \text{ MHz}, \quad \lambda = 0.345 \text{ m}$$

$$d = 3.7 \text{ km}$$

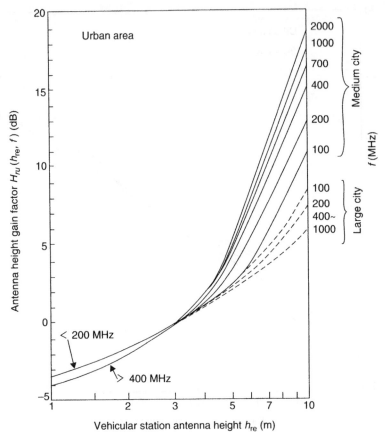

Figure 7.11   Plot of H_{ru}, the mobile station height correction factor for the Okumura model. (Figure 4.9 from Ref. 13, courtesy of Wiley.)

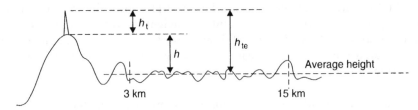

Figure 7.12   Measuring effective transmitter height. (Figure 4.10 from Ref. 13, courtesy of Wiley.)

What is the predicted path loss using the Okumura model?
First, it is readily determined that

$$L_{FS} = 102.6 \text{ dB}$$

Then the required correction factors from Figures 7.9 and 7.10 are incorporated to get the resulting median path loss:

$$L_{50}(\text{dB}) = 102.6 + 26 - (-8) = 136.6 \text{ dB}$$

Note that an H_{ru} correction factor is not required since the mobile antenna is at 3 m, which is the reference height. $\square$

7.4.3 Hata Model

The Hata model (sometimes called the Okumura–Hata model) is an empirical formulation that incorporates the graphical information from the Okumura model [21–23]. There are three different formulas for the Hata model: for urban areas, for suburban areas, and for open areas.

Urban Areas

$$L_{50}(\text{dB}) = 69.55 + 26.16\log(f_c) - 13.82\log(h_t) - a(h_r) + [44.9 - 6.55\log(h_t)]\log(d)$$

$$(7.14)$$

where

$150 < f_c < 1500, f_c$ in MHz
$30 < h_t < 200, h_t$ in m
$1 < d < 20, d$ in km

and $a(h_r)$ is the mobile antenna height correction factor. For a small- or medium-sized city:

$$a(h_r) = (1.1\log(f_c) - 0.7)h_r - (1.56\log(f_c) - 0.8), \qquad 1 \le h_r \le 10 \text{ m} \quad (7.15)$$

and for a large city:

$$a(h_r) = \begin{cases} 8.29(\log(1.54h_r))^2 - 1.1, & f_c \le 200 \text{ MHz} \\ 3.2(\log(11.75h_r))^2 - 4.97, & f_c \le 400 \text{ MHz} \end{cases} \qquad (7.16)$$

Suburban Areas

$$L_{50}(\text{dB}) = L_{50}(\text{urban}) - 4.78(\log(f_c))^2 + 18.33\log(f_c) - 40.94 \qquad (7.17)$$

Open Areas

$$L_{50}(\text{dB}) = L_{50}(\text{urban}) - 2\left(\log\left(\frac{f_c}{28}\right)\right)^2 - 5.4 \qquad (7.18)$$

The Hata formulation makes the Okumura model much easier to use and is usually the way the Okumura model is applied.

Example 7.5. Consider the same system used in Example 7.4. Determine the median path loss using the Hata model.

$$h_t = 68 \text{ m}, \qquad f = 870 \text{ MHz}, \qquad \lambda = 0.345 \text{ m}$$
$$h_r = 3 \text{ m}, \qquad d = 3.7 \text{ km}$$

Then

$$L_{50}(\text{dB}) = 69.55 + 26.16\log(870) - 13.82\log(68) - a(h_r)$$
$$+ [44.9 - 6.55\log(68)]\log(3.7)$$

where the mobile antenna height correction factor (assuming a large city) is

$$a(3) = 3.2(\log(11.75 \cdot 3))^2 - 4.97 = 2.69$$

The final result is then $L_{50}(\text{dB}) = 137.1$ dB, which is in agreement with Example 7.4. □

7.4.4 COST 231 Model

The COST 231 model, sometimes called the Hata model PCS extension, is an enhanced version of the Hata model that includes 1800–1900 MHz [22]. While the Okumura data extends to 1920 MHz, the Hata model is only valid from 150 to 1500 MHz. The COST 231 model is valid between 1500 and 2000 MHz. The coverage for the COST 231 model is [23]

Frequency: 1500–2000 MHz
Transmitter (base station) effective antenna height, h_{te}: 30–200 m
Receiver (mobile) effective antenna height, h_{re}: 1–10 m
Link distance, d: 1–20 km

The COST 231 median path loss is given by

$$L_{50}(\text{dB}) = 46.3 + 33.9\log(f_c) - 13.82\log(h_t) - a(h_r)$$
$$+ [44.9 - 6.55\log(h_t)]\log(d) + C \qquad (7.19)$$

where

> f_c is the frequency in MHz
> h_t is the base station height in meters
> h_r is the mobile station height in meters
> $a(h_r)$ is the mobile antenna height correction factor defined earlier
> d is the link distance in km
> $C = 0\,\text{dB}$ for medium cities or suburban centers with medium tree density
> $C = 3\,\text{dB}$ for metropolitan centers

The COST 231 model is restricted to applications where the base station antenna is above the adjacent roof tops. Hata and COST 231 are central to most commercial RF planning tools for mobile telephony.

7.4.5   Lee Model

The Lee model [24–26] was originally developed for use at 900 MHz and has two modes: *area-to-area* and *point-to-point*. Even though the original data are somewhat restrictive in its frequency range, the straightforward implementation, ability to be fitted to empirical data, and the results it provides make it an attractive option. The model includes a frequency adjustment factor that can be used to increase the frequency range analytically. The Lee model is a modified power law model with correction factors for antenna heights and frequency. A typical application involves taking measurements of the path loss in the target region and then adjusting the Lee model parameters to fit the model to the measured data.

Lee Area-to-Area Mode   For area-to-area prediction, Lee uses a reference median path loss at one mile, called L_0, the slope of the path loss curve, γ in dB/decade, and an adjustment factor F_0. The median loss at distance, d, is given by

$$L_{50}(\text{dB}) = L_0 + \gamma\log(d) - 10\log(F_0) \qquad (7.20)$$

Lee's model was originally formulated as a received signal level prediction based on a known transmit power level and antenna gains. The formulation presented here has been converted from an RSL model to a path loss model to better fit the format of the other models presented. This means that the

TABLE 7.2   Reference Median Path Loss for Lee's Model

Environment	L_0 (dB)	γ
Free space	85	20
Open (rural) space	89	43.5
Suburban	101.7	38.5
Urban areas		
Philadelphia	110	36.8
Newark	104	43.1
Tokyo	124.0	30.5

Source: Derived from Ref. 26, with L_0 values adjusted to 1 km.

power adjustment factor from the original Lee model is not required since the path loss is independent of the transmit power. In addition, the reference path loss distance has been modified from Lee's original value at one mile to the corresponding value at 1 km. The slope of the path loss curve, γ, is the exponent of the power law portion of the loss (expressed as a dB multiplier). Some empirical values for the reference median path loss at 1 km and the slope of the path loss curve are given in Table 7.2. Data for any given application will deviate from these data, but should be of the same order of magnitude.

The basic setup for collecting this information is as follows:

$$f = 900\,\text{MHz}$$
$$G_b = 6\,\text{dBd} = 8.14\,\text{dBi}$$
$$G_m = 0\,\text{dBd} = 2.14\,\text{dBi}$$

To see how the L_0 are computed, first consider the free-space case:

$$L_0 = 20\log\left(\frac{\sqrt{G_bG_m}\,\lambda}{4\pi d}\right)$$

or, in dB,

$$L_0 = G_b + G_m - 22 + 20\log(\lambda) - 20\log(d)$$

where λ and d are in the same units. Substituting the appropriate values from above and using $d = 1000\,\text{m}$ yields

$$L_0 = -81.2\,\text{dB}$$

Lee's empirical data suggests that $L_0 = -85\,\text{dB}$, which is likely a result of the antennas not being ideal or the test not being ideally free space. Using the P_{r0}

values from Ref. 26 to determine the L_0 values is also straightforward. The measured P_0 values given by Lee were measured using the above conditions and a 10-W transmitter. Thus the computation is

$$L_0(\text{dB}) = P_0(\text{dBm}) - 40 \text{ dBm}$$

For free space, Lee measured $P_0 = -45\,\text{dBm}$, which gives $L_0 = -85\,\text{dB}$ as stated above. In Newark, Lee measured $P_0 = -64\,\text{dBm}$, so $L_0 = -104$.

The adjustment factor, F_0, is comprised of several factors, $F_0 = F_1 F_2 F_3 F_4 F_5$, which allow the user to adjust the model for the desired configuration. Note that the numbering of these factors is not universal.

The *base station antenna height* correction factor is

$$F_1 = \left(h_b(\text{m})/30.48\right)^2 = \left(h_b(\text{ft})/100\right)^2$$

The *base station antenna gain* correction factor is

$$F_2 = \left(G_b/4\right)$$

where G_b is the actual base station antenna gain relative to a half-wave dipole.

The *mobile antenna height* correction factor is

$$F_3 = \left(h_m(\text{m})/3\right)^2 \qquad \text{if } h_m(\text{m}) > 3$$
$$F_3 = \left(h_m(\text{m})/3\right) \qquad \text{if } h_m(\text{m}) < 3$$

The *frequency* adjustment factor is

$$F_4 = \left(f/900\right)^{-n}, \qquad \text{where } 2 < n < 3 \text{ and } f \text{ is in MHz}$$

The *mobile antenna gain* correction factor is

$$F_5 = G_m/1$$

where G_m is the gain of the mobile antenna *relative to a half-wave dipole*.

For these correction factors, it is important to recognize that misprints in the signs of the correction factors can sometimes be found in the literature. Such errors can result in confusion and invalid results if not recognized. The best advice is to apply a simple test case if in doubt.

Lee Point-to-Point Mode The point-to-point mode of the Lee model includes an adjustment for terrain slope. The median path loss is given by

$$L'_{50}(\text{dB}) = L_{50}(\text{dB}) - 20\log\left(\frac{h_{\text{eff}}}{30}\right) \qquad (7.21\text{a})$$

or

$$L_{50}(\text{dB}) = L_0 + \gamma \log(d) - 10\log(F_0) - 20\log\left(\frac{h_{\text{eff}}}{30}\right) \qquad (7.21b)$$

where h_{eff} is in meters. h_{eff} is determined by extrapolating the terrain slope at the mobile back to the base station antenna and then computing the antenna height (vertically) above the extrapolated line see Figure 7.13. The sign of the h_{eff} term is another place where typographical errors can sometimes be found.

Lee indicates that the standard deviation of the error in the area-to-area mode is 8 dB [28] and that for the point-to-point mode is 3 dB [29]. The frequency adjustment coefficient for F_4 is $n = 2$ for suburban or open areas with $f < 450\,\text{MHz}$ and $n = 3$ for urban areas and $f > 450\,\text{MHz}$ [26]. Other cases must be determined empirically.

Example 7.6. What is the expected path loss for a mobile communication system operating at 600 MHz over suburban terrain, for path lengths between 1 and 5 km? The base station antenna is a 5-dBi colinear antenna at 20-m height, and the mobile antenna is a quarter-wave vertical with 0-dBi gain at 1-m height.

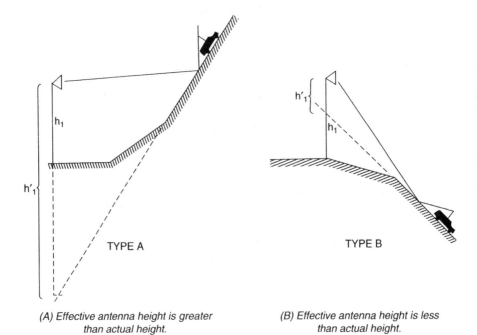

(A) Effective antenna height is greater
than actual height.

(B) Effective antenna height is less
than actual height.

Figure 7.13 Determination of the effective base station antenna height for the Lee model point-to-point mode. (Figure 2.15 from Ref. 27, courtesy of Wiley.)

Since this is a mobile system, the area mode of the Lee model is used. From Table 7.2 an appropriate value of L_0 is

$$L_0 = -101.7 \text{ dB}$$

and

$$\gamma = 38.5$$

The adjustment factors are

$$F_1 = (h_b(\text{m})/30.48)^2 = (20/30.48)^2 = 0.431$$
$$F_2 = (G_b/4) = 3.2/4 = 0.791$$
$$F_3 = (h_m(\text{m})/3) = 1/3 \qquad \text{since } h_m(\text{m}) < 3$$
$$F_4 = (600/900)^{-n} = 2.76$$
$$F_5 = 1$$

where a value of 2.5 was assumed for n. The compilation of these terms results in

$$F_0 = -5.0 \text{ dB}$$

So the median path loss for this system is given by

$$L_{50} = 106.7 + 38.5\log(d) \text{ dB}$$

where d is expressed in kilometers. Figure 7.14 shows the resulting median path loss along with the corresponding free-space loss for the same distance at 600 MHz. □

It is important to remember the conditions that were used to collect the data. For instance, if a different gain base station antenna is used for the data collection, then the equation for the F_2 correction factor will need to be modified accordingly before using the model. If a simple dipole were used, then the correction factor would simply be

$$F_2 = G_b \text{ relative to a dipole}$$

7.4.6   Comparison of Propagation Models for Built-Up Areas

Table 7.3 provides a high-level comparison of the propagation models discussed in this section. This is, of course, not a complete list of models, but it is a list of the models covered in this chapter and represents several of the more popular models in use today. From this table, it is clear that for applications

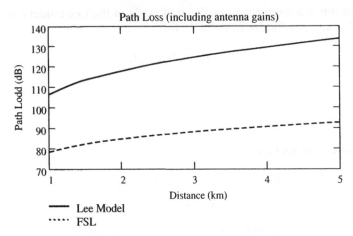

Figure 7.14 Path loss from Lee model for Example 7.6, with free-space loss shown for reference.

TABLE 7.3 Comparison of Propagation Models for Built-Up Areas

Model	Application	Frequency (MHz)	Advantages	Disadvantages
Young	Power law with beta factor	150–3700	Easily applied	Limited data, NYC 1952 only
Okumura	Equation with correction factors from plots	200–1920	Widely used as a reference	Limited data, Tokyo 1960, tedious to apply
Hata	Equation	150–1500	Widely used, straightforward to apply	Based on limited data, does not cover PCS band
COST 231	Equation	1500–2000	Same as Hata but also covers PCS frequencies	
Lee	Equation with computed correction factors	900, plus analytic extension	Relatively easy to apply, can be fitted to measurements, two modes	Requires local data collection for good accuracy

outside of the personal communications bands, the Lee model is going to be the most popular choice. The fact that the Lee model can be fitted to such a wide variety of scenarios makes it a sound choice as well.

7.5 SUMMARY

Many applications in RF and wireless involve propagation of electromagnetic waves in close proximity of the earth's surface. Thus it is important to be able to model the effects of terrain, foliage, and urban structures. The foliage models presented are Weissberger's model, the early ITU model, and the recent ITU model. Weissberger's model and the early ITU model are both based on a power of the frequency and of the depth of the foliage. The updated ITU model for one terminal in woodland is an exponential model, while the model for other foliage scenarios is a dual-slope model that uses the size of the illuminated foliage area to predict the amount of loss due to the foliage. The updated ITU model includes provisions for limiting the foliage loss to the loss on the diffraction path (i.e., using the lesser of the two losses).

Terrain loss can be easily modeled using the Egli model, which is a fourth-power law with a *clutter factor* multiplier to fit the model to empirical data. The Longley–Rice model is a very mature, well-validated model that has gained wide acceptance over many decades of use. The model takes many factors into account and provides accurate predictions of terrain loss.

Propagation loss in built-up areas has been studied extensively in support of mobile telephony, and many different models are available, with different implementations, applicability, and levels of fidelity. The most widely recognized models are the Okumura and Hata's analytic formulation of the Okumura model. The Okumura model is based on data collected in Tokyo in 1960 and thus may have limited applicability, but its wide following makes it valuable for a first-cut analysis and for comparisons. A similar model that is not quite so well known is the Young model. The Young model is based on measurements taken by Young in New York City and may be more representative of modern urban conditions. The Lee model is a modified power law with several adjustment factors to correct for deviations from the configuration of the baseline. The model can be readily adjusted to accommodate any measurements that are available for the region of interest. The Lee model features both an area mode and a point-to-point mode for fixed link scenarios.

While not exhaustive, the set of models presented in this chapter provide some insight into the nature of available models. In the competitive environment of wireless telecommunications, many organizations have developed proprietary models, which they feel best predict the performance of their products. Most of the commercial telecommunication modeling packages will include several different models. It is important to understand which models are being used and the limitation of those models for the particular application. Use of proprietary models in commercial propagation prediction software is unusual and generally not desirable because the credibility (although not necessarily the accuracy) of a model is proportional to how widely accepted it is.

REFERENCES

1. H. L. Bertoni, et al., UHF propagation prediction for wireless personal communications, *Proceedings of the IEEE*, September 1994, pp. 1333–1359.
2. J. D. Parsons, *The Mobile Radio Propagation Channel*, 2nd ed., Wiley, West Sussex, 2000, pp. 52–53.
3. N. Blaunstein, *Radio Propagation in Cellular Networks*, Artech House, Norwood, MA, 2000, p. 172.
4. J. D. Parsons, *The Mobile Radio Propagation Channel*, 2nd ed., Wiley, West Sussex, 2000, pp. 53–54.
5. ITU-R Recommendations, *Attenuation in vegetation*, ITU-R P.833-3, Geneva, 2001.
6. J. J. Egli, Radio Propagation above 40 MC over irregular terrain, *Proceedings of the IRE*, October 1957.
7. N. Blaunstein, *Radio Propagation in Cellular Networks*, Artech House, Norwood, MA, 2000, pp. 156–157.
8. N. Blaunstein, *Radio Propagation in Cellular Networks*, Artech House, Norwood, MA, 2000, pp. 159–163.
9. J. D. Parsons, *The Mobile Radio Propagation Channel*, 2nd ed., Wiley, West Sussex, 2000, pp. 56–60.
10. T. S. Rappaport, *Wireless Communications, Principles and Practice*, 2nd ed., Prentice-Hall, Upper Saddle River, NJ, 2002, p. 145.
11. Irregular Terrain Model (ITM), from the NTIA web site, http://ntiacsd. ntia.doc.gov/msam/
12. ITU-R Recommendations, *Propagation data and prediction methods required for the design of terrestrial line-of-sight systems*, ITU-R P.530-9, Geneva, 2001.
13. J. D. Parsons, *The Mobile Radio Propagation Channel*, 2nd ed., Wiley, West Sussex, 2000, Chapter 4.
14. J. D. Parsons, *The Mobile Radio Propagation Channel*, 2nd ed., Wiley, West Sussex, 2000, pp. 77–79.
15. N. Blaunstein, *Radio Propagation in Cellular Networks*, Artech House, Norwood, MA, 2000, pp. 254–255.
16. J. D. Parsons, *The Mobile Radio Propagation Channel*, 2nd ed., Wiley, West Sussex, 2000, p. 74.
17. T. S. Rappaport, *Wireless Communications, Principles and Practice*, 2nd ed., Prentice-Hall, Upper Saddle River, NJ, 2002, pp. 150–153.
18. N. Blaunstein, *Radio Propagation in Cellular Networks*, Artech House, Norwood, MA, 2000, pp. 259–261.
19. W. C. Y. Lee, *Mobile Communication Engineering, Theory and Applications*, 2nd ed., McGraw-Hill, New York, 1998, pp. 127–129.

20. W. C. Y. Lee, *Mobile Communication Design Fundamentals*, 2nd ed., Wiley, New York, 1993, p. 68.

21. N. Blaunstein, *Radio Propagation in Cellular Networks*, Artech House, Norwood, MA, 2000, pp. 261–264.

22. J. D. Parsons, *The Mobile Radio Propagation Channel*, 2nd ed., Wiley, West Sussex, 2000, pp. 85–86.

23. T. S. Rappaport, *Wireless Communications, Principles and Practice*, 2nd ed., Prentice-Hall, Upper Saddle River, NJ, 2002, pp. 153–154.

24. N. Blaunstein, *Radio Propagation in Cellular Networks*, Artech House, Norwood, MA, 2000, pp. 275–279.

25. W. C. Y. Lee, *Mobile Communication Engineering, Theory and Applications*, 2nd ed., McGraw-Hill, New York, 1998, pp. 124–126.

26. W. C. Y. Lee, *Mobile Communication Design Fundamentals*, 2nd ed., Wiley, New York, 1993, pp. 59–67.

27. W. C. Y. Lee, *Mobile Communication Design Fundamentals*, 2nd ed., Wiley, New York, 1993, p. 74.

28. W. C. Y. Lee, *Mobile Communication Design Fundamentals*, 2nd ed., Wiley, New York, 1993, p. 51.

29. W. C. Y. Lee, *Mobile Communication Design Fundamentals*, 2nd ed., Wiley, New York, 1993, p. 88.

EXERCISES

1. What is the expected foliage loss for a 10-GHz communication system that must penetrate 18 m of foliage?

(a) Using the Wiessberger model

(b) Using the early ITU model

2. How much foliage attenuation is expected for an 800-MHz communication system that must penetrate up to 40 m of foliage?

3. For a ground-based communication system operating at 1.2 GHz, with one terminal located 100 m inside of a wooded area, what is the predicted foliage loss from the updated ITU model? Assume the antenna gains are each 3 dB and the antennas are vertically polarized.

4. Use the Egli model to determine the median path loss for a 400-MHz system over a 5-km path if both antennas are handheld ($h \sim 1.5\,\mathrm{m}$)?

5. Repeat problem 4, but compute the 90% path loss (i.e., what level of path loss will not be exceeded 90% of the time?)

6. Use the Okumura model to predict the median path loss for a 900 MHz system at 10 km in an urban environment. Assume that the mobile antenna height is 7 m and the base station antenna height is 50 m.

7. Use the Hata–Okumura model to determine the expected path loss for a 1-km path in a large city for a 1-GHz system. The receive (mobile) antenna is at 7-m height and the transmit antenna is at 35-m height. You may want to compute the free-space loss for the same geometry to provide a sanity check for your answer.

8. Use the extended COST 231–Hata model to determine the maximum cell radius for a 1.8-GHz system in a medium-sized city ($C = 0\,dB$) if $h_t = 75\,m$ and $h_r = 3\,m$. Assume that the allowable path loss is 130 dB.

9. Use the Lee model to determine the maximum cell radius for a 900-MHz system in a suburban area. Assume that $h_r = 7\,m$ and $h_t = 50\,m$ and that the allowable path loss is 125 dB. You may assume that the mobile antenna gain is at the reference value (0 dBd) and the transmitter antenna gain is also at the reference value of 6 dBd.

CHAPTER 8

Fading and Multipath Characterization

8.1 INTRODUCTION

In this chapter, fading due to shadowing, blockage, and multipath is discussed. In most contexts, the term *fading* is applied to signal loss that changes fairly slowly relative to the signal bandwidth and the term *scintillation* is used to describe rapid variations in signal strength. These terms are usually applied to atmospheric phenomena, however. When modeling a terrestrial mobile radio channel, the principal effects are usually due to terrain and terrain features (including urban features). In this context, it is customary to talk about fading in terms of *large-scale* or *small-scale* fading rather than scintillation. Small-scale fading is further characterized as fast or slow and as spectrally flat or frequency-selective.

Fading is roughly grouped into two categories: large-scale and small-scale fading. Large-scale fading is sometimes called slow fading or shadowing, although the term *slow fading* has a more precise definition in the context of small-scale fading. Large-scale fading is often characterized by a log-normal probability density function (pdf) and is attributed to shadowing, and the resulting diffraction and/or multipath. Changes in large-scale fading are associated with significant changes in the transmitter/receiver geometry, such as when changing location while driving.

Small-scale fading is associated with very small changes in the transmitter/receiver geometry, on the order of a wavelength. Small-scale fading may be either fast or slow and is due to changes in multipath geometry and/or Doppler shift from changes in velocity or the channel. Thus the changes can occur quickly and frequently. Small-scale fading is generally characterized by a Rayleigh or Rician probability density function and is due almost exclusively to multipath.

The multipath discussed in this chapter is due to terrain features and does not include atmospheric multipath, which was treated in Chapter 6. In addition to increasing path loss (fading), multipath may also degrade signal quality. The effects may be temporal or spectral. Temporal effects occur as the delay

Introduction to RF Propagation, by John S. Seybold

from the different paths causes a (time) smeared version of the signal to be received, which is called *delay spread*. Spectral effects occur as changes in the geometry or environment cause Doppler (spectral) shift, which smears the signal spectrum. This effect is called *Doppler spread*.

In most environments, the path loss is sufficiently variable that it must be characterized statistically. This is particularly true for mobile communications where either or both terminals may be moving (changing the relative geometry) and where both are using wide-angle or omnidirection antennas. Multipath models vary depending upon the type of environment and the frequencies involved. While detailed databases of most urban areas are available, statistical modeling based on empirical data (oftentimes fitted to specific empirical data) is still the method of choice. One exception is the planning of fixed line-of-sight (LOS) links, which can be facilitated by geometric multipath computations using urban databases.

The analysis/prediction of channel path loss in a terrestrial mobile environment is broken into three parts:

- First, determine the median (or sometimes mean) path loss based on a model appropriate for the local environment using a model such as the Hata or Lee models, free-space loss, or ground-bounce propagation.
- Second, the large-scale variation of the path loss based on terrain variations and displacement of either the transmitter or receiver is determined. This is referred to as *shadowing* and is usually modeled by a log-normal probability density function.
- Third, the short-term variations about the median due to changes in the multipath and/or Doppler profile of the channel are characterized. This is called *small-scale fading* and may be characterized as fast or slow and time-dispersive or frequency-dispersive.

Before looking at the details of large- and small-scale fading, the impact of ground-bounce multipath and diffraction are addressed.

8.2 GROUND-BOUNCE MULTIPATH

Most terrestrial communication systems do not operate in a free-space environment, but rather must account for the effect of the earth's surface on the propagation path. There are two key effects: ground reflection and path blockage and/or diffraction when part of the path is beyond line of sight. This section covers the effect of a reflective earth on a near-earth propagation path. When the propagation path is near the earth's surface and parallel to it, severe fading can occur if the ground is sufficiently reflective.

Consider a point-to-point communications link operating in close proximity to the earth's surface as shown in Figure 8.1. For this analysis, a flat, smooth,

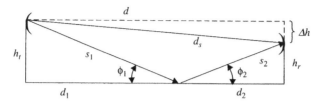

Figure 8.1 Link geometry for ground-bounce reflection.

and reflective ground surface is assumed. Thus there will be "specular" ground reflection from a "flat earth." Specular reflection occurs if and only if the angle of incidence equals the angle of departure at the reflection point (i.e., $\phi_1 = \phi_2$).

Referring to Figure 8.1, the following observations can be made:

$$\phi_1 = \phi_2$$

The slant range is

$$d_s = \sqrt{d^2 + \Delta h^2}$$

and

$$d = d_1 + d_2 \tag{8.1}$$

Expressions for the reflection angles are as follows:

$$\phi_1 = \tan^{-1}\left(\frac{h_t}{d_1}\right), \qquad \phi_2 = \tan^{-1}\left(\frac{h_r}{d_2}\right)$$

From these expressions and the fact that the two angles are equal, it is clear that

$$\frac{h_t}{d_1} = \frac{h_r}{d_2}$$

Next, using (8.1), the following equation can be written:

$$d_1 h_r = (d - d_1)h_t$$

Solving this equation for d_1 yields

$$d_1 = \frac{d h_t}{h_r + h_t}$$

Replacing d_1 by $d - d_2$ in this equation and solving for d_2, provides the following equations for d_2.

$$d_2 = d - \frac{dh_t}{h_r + h_t}$$

$$d_2 = \frac{dh_r}{h_r + h_t} \tag{8.2}$$

Thus the specular reflection point between the two antennas can be determined by knowing the heights of the antennas. The following equations follow from the specular geometry:

$$s_1 = \sqrt{d_1^2 + h_t^2} = h_t \sqrt{1 + \frac{d^2}{(h_r + h_t)^2}} \tag{8.3}$$

$$s_2 = \sqrt{d_2^2 + h_r^2} = h_r \sqrt{1 + \frac{d^2}{(h_r + h_t)^2}} \tag{8.4}$$

And also from the geometry, the slant range is found to be

$$d_s = \sqrt{d^2 + (h_t - h_r)^2} \tag{8.5}$$

The received signal can be found by taking the vector sum of the direct and reflected wave at the receive antenna. The magnitude and phase of the reflected signal are determined by the path-length difference (primarily phase since the distances are nearly equal) and the *reflection coefficient* of the ground. The reflection coefficient, often denoted by ρ, is a complex parameter that modifies both the magnitude and phase of the reflected wave.

The curves in Figure 8.2 and Figure 8.3 [1] indicate that for a smooth, reflective surface, the magnitude of the reflection coefficient is approximately one, regardless of the polarization and frequency, if the grazing angle is small. Figure 8.4 shows that the phase angle of the reflection coefficient is near 180 degrees for vertical polarization. Not shown in the figures is the phase angle for horizontal polarization, which is virtually 180 degrees regardless of the frequency and grazing angle. Thus for the ground-bounce case the reflection coefficient is well-approximated by $\rho = -1$ as long as the surface is smooth and conductive and the angle of incidence is small.

The relative phase of the reflected signal when it reaches the receiver must be computed since the relative phase determines how the direct and reflected signals will combine. The reflection coefficient of the ground depends upon four factors:

- Angle of incidence (assumed to be very small)
- Ground material properties (flat, smooth, and conductive)

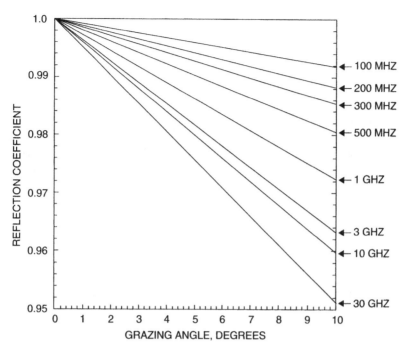

Figure 8.2 Reflection coefficient magnitude versus grazing angle for seawater using horizontal polarization. (Figure 6.7 from Ref. 1, courtesy of Munro Publishing Company.)

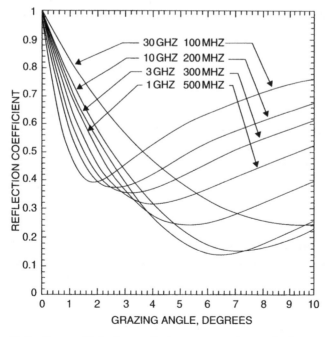

Figure 8.3 Reflection coefficient magnitude versus grazing angle for seawater using vertical polarization. (Figure 6.8 from Ref. 1, courtesy of Munro Publishing Company.)

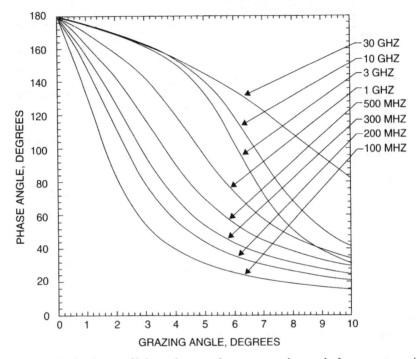

Figure 8.4 Reflection coefficient phase angle versus grazing angle for seawater using vertical polarization. (Figure 6.9 from Ref. 1, courtesy of Munro Publishing Company.)

- Frequency
- Polarization (if the grazing angle is small enough, the polarization does not affect the reflection)

Since the grazing angle is assumed to be small, the reflection undergoes a 180-degree phase shift at the point of reflection regardless of polarization. The small grazing angle also means that $|\rho| \sim 1$.

For geometries where $d_s \gg h_t$ and $d_s \gg h_r$ (i.e., ϕ is very small), the following approximation can be made:

$$s_1 + s_2 \sim d_s$$

Thus the path loss (free-space loss) is approximately the same for the direct and reflected paths. In this case, if the waves are exactly in phase, there will be a 6-dB increase in the received signal (twice the amplitude implies four time the power). If the waves are exactly 180 degrees out of phase, the signal will cancel completely and no signal will be received.

The **E** field at the receiver can be expressed as

$$\mathbf{E} = E_d + E_d \rho e^{-j\Delta\theta}$$

where

E is the electric field at the receiver

E_d is the electric field due to the direct wave at the receiver

$\Delta\theta$ is the phase difference between the direct and reflected wave fronts due to the path-length difference

For the small grazing angle case, $\rho = -1$ and

$$\mathbf{E} = E_d - E_d e^{-j\Delta\theta}$$

$$\mathbf{E} = E_d(1 - e^{-j\Delta\theta})$$

The magnitude of the **E** field can be expressed as

$$|\mathbf{E}| = |E_d[1 - \cos(\Delta\theta) + j\sin(\Delta\theta)]|$$

The expression for the magnitude of **E** can be expanded as follows:

$$|\mathbf{E}| = |E_d|\{[1 - \cos(\Delta\theta) + j\sin(\Delta\theta)] \cdot [1 - \cos(\Delta\theta) - j\sin(\Delta\theta)]\}^{1/2}$$

$$|\mathbf{E}| = |E_d| \left[1 - 2\cos(\Delta\theta) + \underbrace{\cos^2(\Delta\theta) + \sin^2(\Delta\theta)}_{1} \right]^{1/2}$$

$$|\mathbf{E}| = |E_d|[2 - 2\cos(\Delta\theta)]^{1/2}$$

$$|\mathbf{E}| = |E_d|2 \underbrace{\sqrt{\frac{1}{2} - \frac{1}{2}\cos(\Delta\theta)}}_{\sin(\Delta\theta/2)}$$

Thus over a flat, reflective surface, the magnitude of **E** field at the receiver can be accurately expressed as

$$|\mathbf{E}| = |E_d|2\sin(\Delta\theta/2) \tag{8.6}$$

where $\Delta\theta$ is the phase difference due to path-length difference and E_d is the **E** field due to the direct return only.

The phase difference between the direct and the reflected wave (due to path-length difference) is

$$\Delta\theta = \frac{s_1 + s_2 - d_s}{\lambda} 2\pi \tag{8.7}$$

From (8.3) and (8.4) it is clear that

$$s_1 + s_2 = (h_t + h_r)\sqrt{1 + \frac{d^2}{(h_r + h_t)^2}}$$

Substituting this result and (8.5) into (8.7) yields the exact expression for the phase difference:

$$\Delta\theta = d\left(\sqrt{1 + \frac{(h_r + h_t)^2}{d^2}} - \sqrt{1 + \frac{(h_t - h_r)^2}{d^2}}\right)\frac{2\pi}{\lambda} \qquad (8.8)$$

which can be used in (8.6) to generate an expression for the exact path loss in the presence of multipath as a function of the free-space loss:

$$\boxed{L_{mp} = L_{fsl}4\sin^2(\Delta\theta/2)} \qquad (8.9)$$

An alternate expression for $\Delta\theta$ can be derived by using the binomial expansion for the square roots. Recall that the binomial expansion is given by

$$\sqrt{1+x} = 1 + \frac{x}{2} - \frac{x^2}{8} + \frac{x^3}{16} - \frac{x^4}{128} + \dots$$

Using

$$x = \left(\frac{h_r + h_t}{d}\right)^2 \quad \text{and} \quad x = \left(\frac{h_r - h_t}{d}\right)^2$$

in the binomial expansion, it is clear that if $d \gg h_r$ and $d \gg h_t$, then

$$\left[\left(\frac{h_r + h_t}{d}\right)^2\right]^2 \approx 0 \quad \text{and} \quad \left[\left(\frac{h_r - h_t}{d}\right)^2\right]^2 \approx 0$$

Thus using only the first two terms of the binomial expansion will be a good approximation. Making the designated substitutions in (8.7), yields

$$\Delta\theta \cong \frac{d}{2}\left(\frac{(h_r^2 + 2h_r h_t + h_t^2) - (h_r^2 - 2h_r h_t + h_t^2)}{d^2}\right)\frac{2\pi}{\lambda}$$

which simplifies to

$$\Delta\theta \cong \frac{2h_r h_t}{d}\frac{2\pi}{\lambda} \qquad (8.10)$$

Using this result for $\Delta\theta$ in (8.6) yields

$$|\mathbf{E}| = |E_d| 2 \sin\left(\frac{2\pi h_t h_r}{d\lambda}\right) \tag{8.11}$$

which indicates that the received power is proportional to the magnitude squared of the **E** field:

$$P_r \propto |\mathbf{E}|^2$$

Therefore, the received signal power will be

$$P_r = 4\frac{|E_d|^2}{\eta}\sin^2\left(\frac{2\pi h_t h_r}{d\lambda}\right) \tag{8.12}$$

where η is the characteritic impedance of free space. But, from the free-space loss equation, the magnitude squared of the direct path **E** field at the receiver, E_d, is given by

$$|E_d|^2 = \eta\frac{P_T G_T G_R \lambda^2}{(4\pi d)^2}$$

So

$$P_r = 4\frac{P_T G_T G_R \lambda^2}{(4\pi d)^2}\sin^2\left(\frac{2\pi h_t h_r}{d\lambda}\right)$$

For this analysis, it was assumed that $d \gg h_t$ and $d \gg h_r$; if it is also true that $d\lambda \gg h_t h_r$, then the following approximation may be made:

$$\sin^2\left(\frac{2\pi h_t h_r}{d\lambda}\right) \cong \left(\frac{2\pi h_t h_r}{d\lambda}\right)^2$$

and

$$P_r \cong 4\frac{P_T G_T G_R \lambda^2}{(4\pi d)^2}\left(\frac{2\pi h_t h_r}{d\lambda}\right)^2$$

which results in the following (approximate) expression for the path loss on a near-earth propagation path over a flat, smooth conducting surface:

$$\boxed{L_{mp} \cong G_T G_R \frac{(h_t h_r)^2}{d^4}} \tag{8.13}$$

This equation indicates that as $d\lambda$ gets large compared to the heights, h_t and h_r, the received power drops off as d^4 rather than d^2 when there is a strong specular reflection with a 180-degree phase shift. This is a worst-case analysis, which provides a good upper bound on the predicted path loss. It is noteworthy that this path-loss expression is independent of the wavelength.

The exact ground-bounce expression is seldom used since the geometry is rarely precise enough to locate the peaks and nulls accurately. The recommended approach is to compute the free-space loss and the approximate ground-bounce path loss and use whichever gives greater loss.

$$\frac{P_R}{P_T} = \min\left(G_T G_R \frac{(h_t h_r)^2}{d^4}, \frac{G_T G_R \lambda^2}{(4\pi d)^2} \right) \tag{8.14}$$

The *crossover point* is defined as the distance at which the $1/d^4$ approximation and free-space loss are equal. It is sometimes helpful to determine where the crossover point occurs and then use the approximation for ranges greater than the crossover range and free-space loss otherwise. The crossover point is found by equating the approximate ground-bounce path loss (8.13), and the free-space loss and solving for d.

$$\frac{(h_t h_r)^2}{d^4} = \left(\frac{\lambda}{4\pi d}\right)^2$$

$$d_x = \frac{4\pi h_t h_r}{\lambda} \tag{8.15}$$

The computed crossover range for the plot in Figure 8.5 is 12.6 km, which is consistent with the curves shown. By first computing the crossover point, one can immediately determine which path loss expression applies to a given situation.

Example 8.1. Consider the following point-to-point communications link:

$$h_T = h_R = 10 \text{ m}$$
$$d = 4 \text{ km}$$
$$f = 2 \text{ GHz} (\lambda = 0.15 \text{ m})$$

What is the predicted path loss for this link?

Since no antenna gains are given, normalized gains (i.e., gain = 0 dB) for transmit and receive are used. The first step is to compute the free-space loss using antenna gain values of unity:

$$FSL = 20 \log\left(\frac{\lambda}{4\pi d}\right)$$

$$FSL = -110.5 \text{ dB}$$

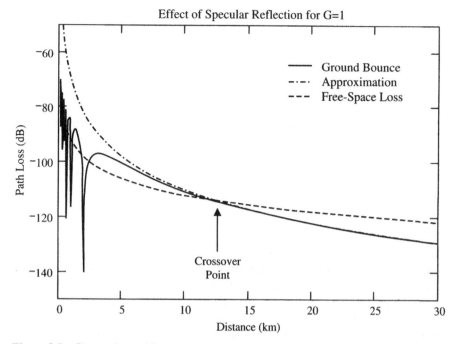

Figure 8.5 Comparison of free-space loss, ground-bounce loss, and the ground-bounce loss approximation, for $h_t = 100\,\text{m}$, $h_r = 3\,\text{m}$, and $\lambda = 0.3\,\text{m}$.

Since $d \gg h_t$ and $d \gg h_r$ and in fact, $d\lambda \gg h_t h_r$, the $1/d^4$ approximation can be used:

$$L_{mp} \cong \frac{(h_t h_r)^2}{d^4} = 3.91 \times 10^{-11} = -104.1\,\text{dB}$$

So the approximated path loss is less loss than free-space loss, and the free-space result should be used: $PL = -110.5\,\text{dB}$.

If instead of 4 km, d is 40 km, then

$$FSL = -130.5\,\text{dB}$$

and

$$L_{mp} = -144.1\,\text{dB}$$

So in this case, the path loss from the $(h_t h_r)^2/d^4$ approximation would be used.

□

In summary, when operating over a flat reflective surface, the path loss may be more or less than that predicted by free-space loss. The 6-dB improvement cannot be ensured unless operating under very controlled conditions, such as a radar instrumentation range. For large distances, the path loss is proportional to $(h_t h_r)^2/d^4$ and is not a function of the wavelength. One may compute the loss both ways and use the method that predicts the greatest path loss or determine the crossover point and use free-space loss for ranges less than the crossover range and the $(h_t h_r)^2/d^4$ approximation for ranges greater than the crossover range.

A similar expression for the ground-bounce path loss over a curved reflective surface can also be developed. One such derivation is presented in Ref. 2. The curved reflective formulation applies to links that operate near the earth's surface and close to the horizon. For most applications, the flat reflective surface formulation is adequate.

8.2.1 Surface Roughness

When determining if ground reflection is likely to be significant, a means of quantifying the smoothness (flatness) of the reflecting surface is required. The *Rayleigh criterion* provides a metric of surface roughness. The Rayleigh roughness is derived based on the terrain variation (Δh) that will provide a 90-degree phase shift at the receiver between a reflection at a terrain peak versus a reflection from a terrain valley at the same distance (see Figure 8.6). The Rayleigh criterion is given by [3]

$$H_R = \frac{\lambda}{8\sin(\theta)} \qquad (8.16)$$

where

λ is the wavelength

θ is the grazing angle of the reflection

From the geometry shown in Figure 8.6, it can be determined that

$$\sin(\theta) = \frac{h_r + h_t}{d}$$

where

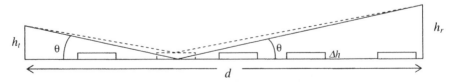

Figure 8.6   Geometry of surface roughness.

h_r is the receiver height

h_t is the transmitter height

d is the communication distance

The Rayleigh criterion can then be expressed as [3]

$$\boxed{H_R = \frac{\lambda d}{8(h_r + h_t)}} \tag{8.17}$$

When the extent of the terrain features, Δh, is less than H_R, the surface can be treated as being smooth. Thus the amount of terrain variation that is tolerable is proportional to the wavelength and the distance traversed and inversely proportional to the height of the antennas above the surface. When a reflective surface is rough, meaning that the extent of the terrain features, Δh, is much larger than H_R, then the reflections from the surface will be diffuse and are characterized as scattering rather than reflection.

8.2.2 Fresnel Zones

Consider the point-to-point link depicted in Figure 8.7. The vector TR is the line of sight between the transmitter and the receiver and the link distance is $d_1 + d_2$.

If there is a diffraction point at P, the signal TPR will combine with TR at R. TPR traverses a slightly greater distance than TR and therefore will have a different phase.

The direct and reflected/diffracted path lengths can be expressed as

$$\text{TR} = d_1 + d_2$$
$$\text{TPR} = \sqrt{d_1^2 + h^2} + \sqrt{d_2^2 + h^2}$$

So the path-length difference is

$$\Delta = \sqrt{d_1^2 + h^2} + \sqrt{d_2^2 + h^2} - d_1 - d_2$$

If $h \ll d_1$, and $h \ll d_2$, then the binomial expansion can be used:

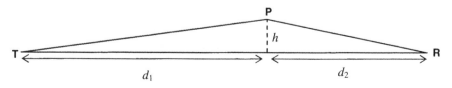

Figure 8.7 Fresnel zone geometry.

$$\sqrt{d_1^2 + h^2} \cong d_1\left(1 + \frac{h^2}{2d_1^2}\right)$$

Making the appropriate substitutions in the equation for Δ, yields

$$\Delta = d_1 + \frac{h^2}{2d_1} + d_2 + \frac{h^2}{2d_2} - d_1 - d_2$$

or

$$\Delta = \frac{h^2}{2d_1} + \frac{h^2}{2d_2}$$

The corresponding phase difference is

$$\phi = \frac{2\pi\Delta}{\lambda}$$

where

$$\Delta = \frac{h^2}{2}\left(\frac{d_1 + d_2}{d_1 d_2}\right)$$

so

$$\phi = \frac{2\pi}{\lambda} \cdot \frac{h^2}{2}\left(\frac{d_1 + d_2}{d_1 d_2}\right) \qquad (8.18)$$

The Fresnel–Kirchhoff diffraction parameter is often used to shorten the notation in Fresnel zone analyses and is defined as

$$v = h\sqrt{\frac{2(d_1 + d_2)}{\lambda d_1 d_2}} \qquad (8.19)$$

If the diffraction point is below the line of sight (LOS), then h is negative and v will also be negative. When the diffraction point is located on the LOS, h and v are both equal to zero. If the blockage is the horizon, then the "h and v equal zero" case corresponds to the maximum LOS distance.

The cases where $\Delta = n\lambda/2$, where n is an integer, can be found by setting $\phi = n\pi$ in (8.18), which yields the following equation:

$$n\lambda = h^2\left(\frac{d_1 + d_2}{d_1 d_2}\right)$$

Thus

$$\phi = \frac{\pi}{2}v^2$$

The destructive reflection/diffraction points can then be identified by defining a term, h_n, such that

$$h_n = \sqrt{\frac{n\lambda d_1 d_2}{d_1 + d_2}} \tag{8.20}$$

Reflectors/diffraction at h_n for odd values of n will cause destructive interference. Since the difference in path lengths is on the order of λ, the reflected/diffracted signal may be as strong as the direct signal and cause cancellation.

The equation for h_n defines a sequence of ellipsoids with the transmit and receive antennas as the foci. Diffractors or reflectors at the odd-numbered Fresnel zone boundaries will cause destructive interference. Figure 8.8 shows a diagram of the Fresnel zones defined by a point-to-point link. Note that this diagram is two-dimensional, whereas the actual Fresnel zones are three-dimensional ellipsoids. For large h or small d_1 and d_2, the antenna pattern may attenuate the undesired signal. For omnidirection (vertical) antennas, there may be attenuation of the undesired signal in elevation, but not in azimuth.

From the preceding analysis, it is clear that any reflectors/diffractors within the field of view should not be near an odd Fresenel zone boundary to avoid signal loss. It is also important that the first Fresnel zone be clear of obstructions because this can seriously degrade the available signal energy. Due to Huygen's principle, covered in the next section, the diffracted electromagnetic energy that fills the shadow at the receive end of the link reduces the energy that arrives at the receiver. If the first Fresnel zone is not clear, then free-space loss does *not* apply and an adjustment term must be included. For most appli-

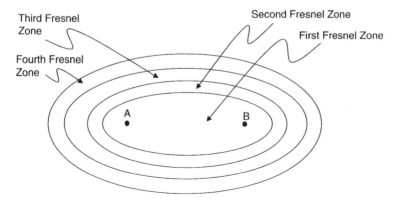

Figure 8.8 Fresnel zones between a transmitter and receiver.

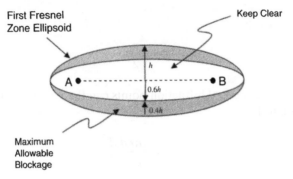

Figure 8.9   Fresnel zone blockage geometry.

cations, having 60% of first Fresnel zone clear is sufficient. The 60% clear applies to the radius of the ellipsoid as shown in Figure 8.9. At the 0.6h point, the Fresnel–Kirchhoff diffraction parameter is $v = -0.8$* and the resulting diffraction loss will be 0 dB [4]. The quantification of diffraction loss is discussed in Section 8.2.4.

Example 8.2.   Consider a point-to-point communication system, with $d = 1$ km and $f = 28$ GHz. If there is a building present, 300 m from one end of the link, how far must it be (in elevation or height) from the LOS to not impede transmission? (That is, find 60% of the first Fresnel zone radius at 300 m.)

   The parameters for determination of the Fresnel zone radius at 300 m are

$$d_1 = 300, \qquad d_2 = 700$$

$$\lambda = 0.107 \text{ m}$$

when eqn (8.20) is applied, the resulting expression for the Fresnel zone radii is:

$$h_n = \sqrt{n \cdot 2.247}$$

   So the rooftop must be at least 0.9 m below the LOS to keep the first Fresnel zone 60% clear. It is also important that the rooftop not be near one of the odd Fresnel zone boundaries to prevent destructive interference from reflections. ☐

It should be emphasized again that reflectors or diffractors near the odd Fresnel zone boundaries produce destructive interference and should be avoided. In addition, blockage within the first Fresnel zone may result in reduced energy at the receiver.

---

*v is negative since the blockage is below the LOS, i.e. the LOS is not obstructed.

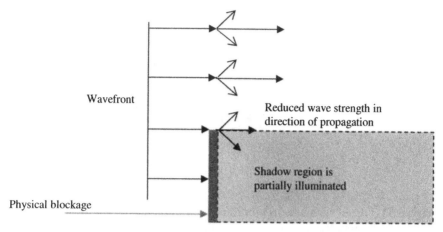

Figure 8.10   Illustration of Huygen's Principle and diffraction.

8.2.3   Diffraction and Huygen's Principle

Diffraction is the physical phenomenon whereby an electromagnetic wave can propagate over or around objects that obscure the line of sight. Diffraction has the effect of filling in shadows, so that some amount of electromagnetic energy will be present in the shadowed region. The easiest way to view the effect of diffraction is in terms of Huygen's Principle. Huygen's Principle states that each point on a wavefront acts as the source of a secondary "wavelet" and all of these "wavelets" combine to produce a new wavefront in the direction of propagation [4–6]. This is illustrated in Figure 8.10.

The "wavelets" from a plane wave generate a plane wave (propagate it) if the extent of the wave front is infinite. If an object blocks part of the wave front, then the "wavelets" near the blockage are not counterbalanced, and radiation in directions other than that of the plane wave will occur. This is diffraction, which causes partial filling of shadows. Regardless of whether the blockage is conductive or nonconductive, the diffraction still occurs. If the blockage is nonconductive, it must not pass any of the RF energy for the diffraction analysis to be valid. The same effect can occur when a signal source is beyond the earth's horizon. While the line of sight is blocked, a small amount of RF energy will be diffracted over the surface and will appear at the receiver.

8.2.4   Quantifying Diffraction Loss

Accurate modeling of losses due to diffraction is very challenging for all but the most basic geometries. One such basic geometry is the knife-edge diffractor, sometimes called a wedge diffractor. This effect has been mathematically characterized and can be used to provide an estimate of the diffraction loss for a single knife edge. The effect of a plane wave incident on a perpendicular conductive barrier can be divided into three shadow regions as shown in

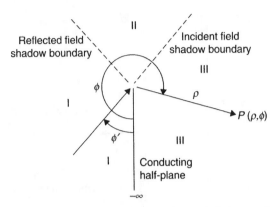

Figure 8.11 Shadow boundaries for diffraction by an infinite conducting half-plane. (Figure 12.12 from Ref. 7, courtesy of Wiley.)

Figure 8.11 [7]. Region I contains direct, reflected, and diffracted rays, region II contains direct and diffracted rays only, and region III contains diffracted rays only. This explains why a knife edge that is below the line of sight may still affect the received wave.

The analysis in this section is based on a conductive barrier. These results are often applied to scenarios where the barriers are not conductive; however, they may not be as precise, depending upon the reflection properties of the barrier. The knife-edge diffracting point may be above, below, or directly on the line of sight between the transmitter and receiver. The electric field due to the diffracted path is given by the diffraction integral,

$$E_d = F(v) = E_0 \frac{(1+j)}{2} \int_0^\infty e^{((-j\pi t^2)/2)} \, dt$$

where E_0 is the electric field at the receiver based on free-space loss only and v is the Fresnel–Kirchhoff diffraction parameter defined earlier. Evaluation of the diffraction integral is usually done numerically or graphically [8]. Lee [9] provides an approximation to the diffraction integral.

$$L_r = 0 \quad \text{dB}, \qquad v \leq -1 \tag{8.21a}$$

$$L_r = 20\log(0.5 - 0.62\,v) \quad \text{dB}, \qquad -1 \leq v \leq 0 \tag{8.21b}$$

$$L_r = 20\log(0.5 e^{-0.95v}) \quad \text{dB}, \qquad 0 \leq v \leq 1 \tag{8.21c}$$

$$L_r = 20\log\left(0.4 - \sqrt{0.1184 - (0.38 - 0.1\,v)^2}\right) \quad \text{dB}, \qquad 1 \leq v \leq 2.4 \tag{8.21d}$$

$$L_r = 20\log\left(\frac{0.225}{v}\right) \quad \text{dB}, \qquad v \geq 2.4 \tag{8.21e}$$

Knife-edge diffraction loss

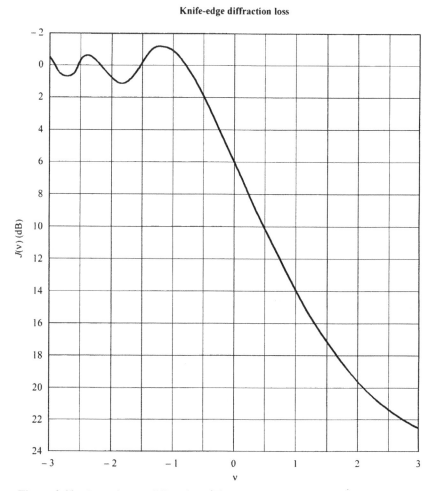

Figure 8.12 Loss due to diffraction. (Figure 7 from Ref. 10, courtesy of ITU.)

Figure 8.12 is a plot of the diffraction loss as a function of the Fresnel–Kirchhoff diffraction parameter, v. Be aware that some authors define the Fresnel–Kirchhoff diffraction parameter with the opposite sign from that used herein. In that case the sign must be reversed to use the plot in Figure 8.12 or incorrect results will be obtained. When v is equal to zero, the first Fresnel zone is 50% blocked and the corresponding signal loss is 6 dB.

Example 8.3. Consider a communication link comprised of two 150-MHz hand-held radios separated by 1 km as shown in Figure 8.13. The barrier between the two radios runs perpendicular to the line of sight and is 5 m below the line of sight. Assume that the barrier is a thin, solid fence that is 200 m from one end of the link. How much additional path loss (beyond free-space loss) can be expected due to the diffraction from the fence?

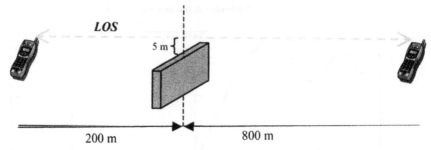

Figure 8.13   Knife-edge diffraction geometry for Example 8.3.

Since the blockage is below the line of sight, the values of h and v are negative. The following parameters are known:

$$h = -5 \text{ m}, \qquad \lambda = 2 \text{ m}$$
$$d_1 = 200 \text{ m}, \qquad d_2 = 800 \text{ m}$$

Using the expression for the Fresnel–Kirchhoff diffraction parameter, (8.19), yields

$$v = -0.395$$

From the plot in Figure 8.12, the diffraction loss can be estimated as 0.75 or -2.5 dB, or the Lee approximation to the diffraction integral can be used directly to get

$$L_r = 20\log(0.5 - 0.62\, v) = -2.6 \text{ dB} \quad \square$$

Unfortunately, most blockages encountered in the real world are not well-modeled by a single knife-edge diffractor. The multiple knife-edge diffraction model given in Ref. 11 can be used, but Rappaport indicates that it tends to oversimplify the geometry and produce somewhat optimistic results. Another option is the rounded-surface diffraction model, where the diffraction is treated as a broadside cylinder as shown in profile in Figure 8.14 [12–15].

The diffraction from a rounded hilltop or surface is determined by computing the knife-edge diffraction for the equivalent height, h, and then computing the excess diffraction loss, L_{ex}, due to the rounded surface. The first step is to determine the radius, r, of the cylinder that circumscribes the actual diffraction points on the obstacle. Then the extent of the diffraction surface, D_S, can be found. The expression for the excess diffraction loss is

$$L_{ex} = -11.7\alpha\sqrt{\frac{\pi r}{\lambda}} \quad \text{dB} \tag{8.22}$$

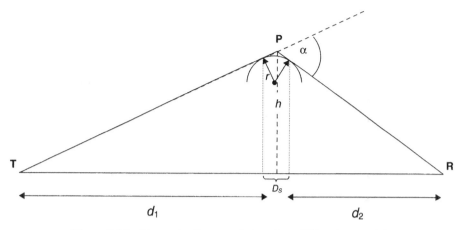

Figure 8.14 Geometry for rounded-surface diffraction model.

where

$$\alpha = v\left[\frac{\lambda(d_1 + d_2)}{2d_1d_2}\right]$$

and r is estimated by

$$r = \frac{2D_s d_1 d_2}{\alpha(d_1^2 + d_2^2)}$$

It is important to note that the excess loss and the knife-edge diffraction loss are actually expressed as gains. That is to say, a negative loss is indeed a loss, not a gain in signal strength.

Example 8.4. Consider a point-to-multipoint communications link operating at 5 GHz over a distance of 1 km shown in Figure 8.15a. There is a pair of narrow hilltops in between the transmitter and receiver, located 300 m from from the transmitter. Find the total diffraction loss if the hilltops are 10 m apart and 3 m above the line of sight.

The individual hilltops are approximated as pointed conductive wedges or knife edges. From the problem statement and the geometry shown in Figure 8.15b, the following observations can be made:

$$BB = 10 \text{ m}, \quad AB = 695 \text{ m}$$
$$ab = 295 \text{ m}, \quad CB = cb = 3 \text{ m}$$

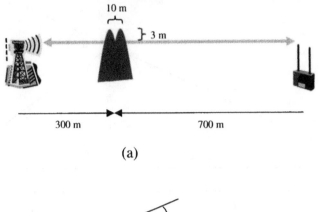

(a)

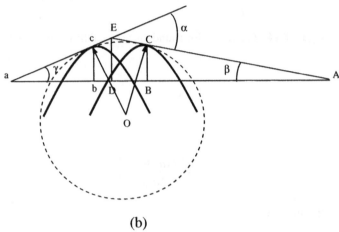

(b)

Figure 8.15 Diffraction geometry for the double-hill blockage problem. (a) Pictorial diagram for Example 8.4. (b) Geometric diagram for Example 8.4.

From this information, the angles β and γ can be found

$$\beta = 0.243°, \qquad \gamma = 0.583°$$

and from basic geometry, the angle, α, is found to be

$$\alpha = \beta + \gamma = 0.826°$$

Now it is possible to determine the value of $h = ED$, by first solving the following set of equations for bD:

$$ED = \tan(\gamma)(ab + bD)$$

$$ED = \tan(\beta)(AB + [10 - bD])$$

Equating these expressions and solving for bD yields

$$bD = \frac{\tan(\beta)(AB+10) - ab\tan(\gamma)}{\tan(\gamma) + \tan(\beta)}$$

or

$$bD = 2.98 \text{ m}$$

from which the value of

$$ED = h = 3.03 \text{ m}$$

can be deduced.

The next step is to determine r, the radius of the circle that is tangential to ae at point c and to AE and point C:

$$r = Oc = OC$$

The interior angle of the circle, $\angle cOC$ can be determined by looking at the interior angles of the quadrangle ceCO. The result is that

$$\angle cOC = \alpha = 0.826°$$

The cord of this angle is

$$cC = bB = 10 \text{ m}$$

Thus cOC is an isosolicles triangle with the equal sides being of length r. Solving for r yields

$$r = \frac{bd/2}{\cos(90 - \alpha/2)} = 694 \text{ m}$$

The one remaining parameter needed to compute the diffraction loss is the Fresnel–Kirchhoff diffraction parameter. Using

$$\lambda = 0.06 \text{ m}$$

$$d_1 = aD = 297.98 \text{ m}$$

$$d_2 = AD = 702.02 \text{ m}$$

and

$$h = 3.03 \text{ m}$$

yields

$$v = 1.2095$$

Thus the basic knife-edge diffraction loss is

$$L_r = 20\log\left(0.4 - \sqrt{0.1184 - (0.38 - 0.1v)^2}\right) = -15.2 \text{ dB}$$

The excess diffraction loss is then determined as

$$L_{ex} = -11.7\alpha\sqrt{\frac{\pi r}{\lambda}} = -32.2 \text{ dB}$$

where α must, of course, be expressed in radians. Therefore the signal at the receiver will be about 47.4 dB lower than it would be for the unobstructed case. This represents a very deep fade, which is to be expected when a 5-GHz signal is blocked by two hilltops. □

For scenarios where the blockage is not well-modeled as a narrow edge of a perfect conductor, the analysis should include the effects of nonideal reflection/diffraction and account for any penetration of the blockage. Such analysis is best treated using the uniform theory of diffraction [16], which is beyond the scope of this text. The UTD permits a variety of shapes, surface textures, and conductivities to be treated. The use of knife-edge diffraction is often sufficient for simple geometries or cases where the specific geometry and material properties are not known with high confidence. When geometry and material specifics are well known, the problem can be treated using the UTD, often in the context of an advanced modeling program. For many RF propagation applications, however, generalized statistical models (such as log-normal shadowing covered in the next section) are sufficient.

According to Lee [14], the diffraction loss will be slightly greater for horizontal polarization than for vertical polarization. Thus the wave in the shadow region for a veritcal polarization signal will be stronger than that of a horizontal polarization signal, all else being equal. In the case of circular polarization, this means that the wave in the shadow region will be elliptically polarized (i.e., the axial ratio will be increased).

8.3 LARGE-SCALE OR LOG-NORMAL FADING

The first step in prediction of the path loss is to determine the median path loss. Several procedures for estimating the median path loss were presented in Chapter 7 and in Section 8.2. The second step is to determine the large-scale fading about that median path loss. Large-scale fading is generally attributable to blockage (diffraction), which is sometimes called *shadowing*. In the case where there is a fair amount of blockage, there may be several diffraction paths to the receiver. These paths are multiplicative, which means that the effects are additive if considered in dB. If there are a sufficient number of diffraction points and/or multireflection paths on the path to the receiver, the Central Limit Theorem may be invoked to justify using a Gaussian random variable to represent the path loss. This is in fact what is frequently done, particularly in cellular telephony. The result is called log-normal fading, since the fading follows a normal distribution in dB (the log domain). For this reason, large-scale fading is modeled as a log-normal random variable. The mean path loss is the same as the median path loss (since for a normal pdf, the mean and median are identical). Thus, only a variance is required to characterize the log-normal fading. This variance is called the *location variability* and is denoted by σ_L.

The probability of a log-normal fade is used to determine the percentage of annular area that is covered at a fixed range from the transmitter. Figure 8.16 shows the variation of RSL with position for a fixed distance, d, as the cell perimeter is traversed. The effects of the normal pdf can be envisioned from this plot.

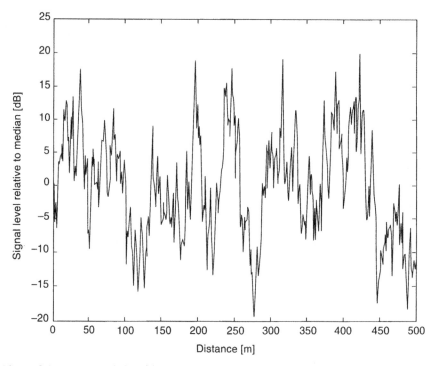

Figure 8.16 Received signal levels relative to the median signal level in a shadowing environment. (Figure 9.2 from Ref. 17, courtesy of Wiley.)

Figure 8.17 shows the histogram and how it compares with the normal pdf. This provides some insight into the validity of log-normal fading models. Figure 8.18 shows a representative plot of path loss measurements. This plot shows the variation of the path loss at any given distance.

Example 8.5. For a given communication system, 90% coverage is desired at the fringe (edge) of the coverage cell. Assume plane earth propagation plus a 20-dB allocation for clutter loss, use $\sigma_L = 6\,dB$ for the shadowing location variability,

$$h_m = 1.5\text{ m}, \qquad h_b = 30\text{ m}$$

and use a maximum allowable path loss of 140 dB. Determine the shadowing fade margin and the distance at which it occurs (i.e., the cell radius). Repeat the problem if $\sigma_L = 8\,dB$.

The expression for the path loss is

$$L(\text{dB}) = -10\log\left(\frac{h_m^2 h_b^2}{d^4}\right) + L_{clutter} + L_s$$

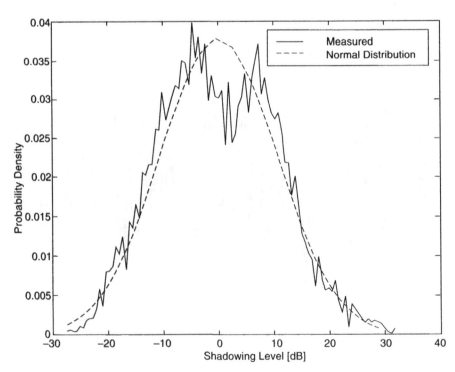

Figure 8.17 Histogram of measured shadowing loss compared to a normal pdf. (Figure 9.3 from Ref. 17, courtesy of Wiley.)

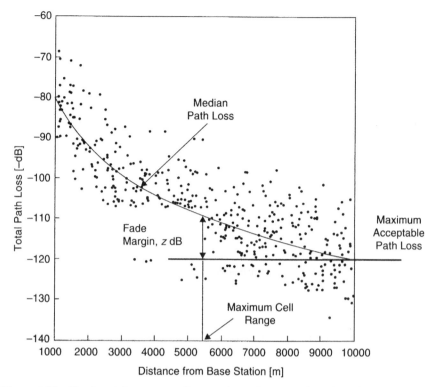

Figure 8.18 Total path loss versus distance from the transmitter in a shadowing environment. (Figure 9.4 from Ref. 17, courtesy of Wiley.)

where L_S is the loss due to shadowing and can be expressed as

$$L_s = z\sigma_L$$

with z being a Gaussian random variable. The probability of an outage is

$$P_{outage} = P(L(\mathrm{dB}) > L_{Max})$$

which can be rewritten as

$$P_{outage} = P(L_S > L_{Max} - L_{PL} - L_{clutter})$$

Since the shadowing pdf is normal and zero mean, the standard normal random variable is

$$z = \frac{L_S}{\sigma_L}$$

which yields

$$P_{outage} = P\left(z > \frac{L_{Max} - L_{PL} - L_{clutter}}{\sigma_L}\right)$$

$$P_{outage} = Q\left(\frac{L_{Max} - L_{PL} - L_{clutter}}{\sigma_L}\right)$$

$$P_{coverage} = 1 - P_{outage} = 1 - Q\left(\frac{L_{Max} - L_{PL} - L_{clutter}}{\sigma_L}\right) = 1 - Q\left(\frac{L_S}{\sigma_L}\right)$$

The net fade margin for the link that can be allocated to shadowing is set equal to the shadowing loss,

$$M = L_S = L_{Max} - (L_{PL}(d) + L_{clutter})$$

and the maximum allowable path loss is

$$L_{PL}(d) = 120 - L_s$$

The shadowing fade margin (based on the probability of a shadowing fade) is

$$L_S = z\sigma_L$$

The value of σ_L is known; and from a Q table (see Table A.1) at $p = 0.1$, z can be found to be 1.28. Thus the required shadowing margin to ensure 90% coverage when $\sigma_L = 6\,dB$ is

$$L_S = 7.7\ dB$$

and the maximum allowable path loss is 112.3 dB, so

$$20\log\left(\frac{h_a h_b}{d^2}\right) = 112.3$$

which yields a cell radius of

$$d = 4.3\ km$$

If instead of $\sigma_L = 6$, $\sigma_L = 8\,dB$ is used, then

$$L_{PL}(d) = 140 - \underbrace{1.28\sigma_L}_{10.24} - 20 = \underbrace{-20\log(h_m h_b)}_{-33.06} + 40\log(d)$$

and

$$d = 3.7 \text{ km} \quad \Box$$

In summary, use $1 - p$, the probability of a shadowing fade, and a Q table to look up z; then z times σ_L, the shadowing location variability, gives the numerical value of L_S, the required shadowing fade margin for the link.

By computing the available link margin as a function of distance, it is possible to generate a curve of the percent coverage versus distance for a given location variability. This is shown in Figure 8.19. Note that the curves all cross at the 50% point as expected since at 50% implies that the probability of being above or below the median is the same regardless of the location variability and the path loss is equal to the median.

Instead of computing the percent coverage at the edge of the cell, one can average over all of the cell area. This can be accomplished by simply computing the coverage over a series of narrow concentric rings and then summing them up [18], where it is implicitly assumed that the coverage area is circular. This is shown conceptually in Figure 8.20. The area of each ring is $2\pi d \Delta d$. Each ring area is multiplied by its corresponding coverage probability, and then they are added and divided by the total area to get the overall cell coverage probability:

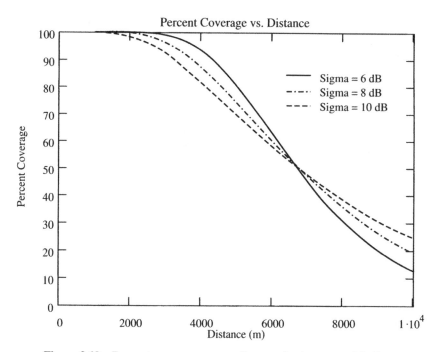

Figure 8.19   Percent coverage versus distance for log-normal fading.

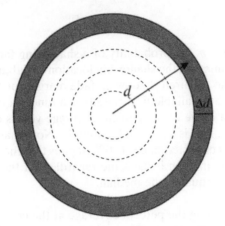

Figure 8.20 Coverage area divided into concentric rings for computing availability over the entire cell.

$$P_{ring}(d) = \left[1 - Q\left(\frac{M(d)}{\sigma_L}\right)\right]$$

$$P_{cell} = \frac{1}{\pi d^2_{max}} \sum P_{ring}(d)(2\pi d)\Delta d$$

As Δd goes to zero, the summation converges to a Reimann sum, which can then be treated as an integral. To avoid confusion with the differential operator, the variable, d, is changed to r:

$$P_{cell} = \frac{1}{\pi d^2_{max}} \int_0^{d_{max}} P_{ring}(r)(2\pi r)\,dr$$

$$P_{cell} = \frac{2}{d^2_{max}} \int_0^{d_{max}} \left[1 - Q\left(\frac{M(r)}{\sigma_L}\right)\right] r\,dr$$

$$P_{cell} = \frac{2}{d^2_{max}} \underbrace{\int_0^{d_{max}} r\,dr}_{\frac{d^2_{max}}{2}} - \int_0^{d_{max}} Q\left(\frac{M(r)}{\sigma_L}\right) r\,dr$$

$$P_{cell} = 1 - \frac{2}{d^2_{max}} \int_0^{d_{max}} rQ\left(\frac{M(r)}{\sigma_L}\right)\,dr \tag{8.23}$$

This equation can be solved numerically for any value of d_{max}. Note that $M(r)$ is a function of r.

The value of σ_L depends on the nature of the terrain. A reasonable fit to the Okumura data is

$$\sigma_L = 0.65(\log(f_c))^2 - 1.3\log(f_c) + A$$

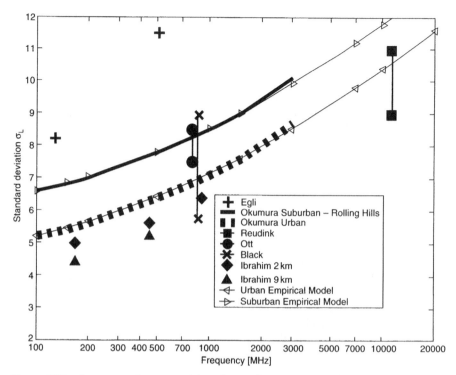

Figure 8.21 Summary of measured location variability values (Figure 9.10 from Ref. 17, courtesy of Wiley.)

where

$A = 5.2$ (urban)

$A = 6.2$ (suburban)

f_c is in MHz

Figure 8.21 from Ref. 17 provides a summary of different measurements of location variability.

8.4 SMALL-SCALE FADING

Small-scale fading encompasses all of the fading that can occur with very small (on the order of one wavelength) changes in the relative position of the transmitter and receiver and, sometimes, reflectors in the environment. The phenomenon of small-scale fading is attributed to the summation of multiple reflected signals arriving with different delays (phases) and amplitudes. The net effect is treated as having a Gaussian random variable for both the in-phase and quadrature components of the received signal. This is equivalent to a Rayleigh (or Ricean) density function for the amplitude and a uniform probability density function for the phase. When the entire signal is comprised of reflected signals, a Rayleigh density function is used. If, on the other hand, there is a single dominant component, such as a line-of-sight path or a large specular reflection in the presence of multiple smaller-strength reflections, a Ricean probability density function is applicable. Appendix A discusses both the Rayleigh and Ricean probability density functions and provides examples of their application.

Small-scale fading is categorized by its spectral properties (flat or frequency-selective) and its rate of variation (fast or slow). The spectral properties of the channel are determined by the amount of delay on the various reflected signals that arrive at the receiver. This effect is called *delay spread* and causes spreading and smearing of the signal in time. The temporal properties of the channel (i.e., the speed of variation) are caused by relative motion in the channel and the concomitant Doppler shift. This is called *Doppler spread* and causes spreading or smearing of the signal spectrum. These parameters are interpreted relative to the signal that is in use; that is, the way a channel is characterized depends upon the relationship between the channel properties and the signal properties. It is important to recognize that the spectral properties of the channel and the rate of variation are independent. This will become clear in the following sections.

8.4.1 Delay Spread

The direct path (if one exists) is the shortest path between the transmitter and receiver. Any multipaths will have traveled greater distances and will therefore be delayed in time relative to the direct signal. Thus the reflected signal(s) will not align with the direct signal, and the cumulative signal will be smeared in time. This is called *delay spread* and is illustrated in Figure 8.22.

There are many metrics of multipath delay spread effects [19]. The *mean excess (average) delay* consists of a weighted average (first moment) of the power delay profile (magnitude-squared of the channel impulse response). The *rms delay spread* is the rms value (second moment) of the power delay profile, denoted by σ_t. The rms delay spread is one of the more widely used characterizations of delay spread and is used with the exponentially decaying impulse response in indoor propagation modeling (Chapter 9). The *delay window* is the width of middle portion of the power delay profile that contains a certain percentage of the energy in the profile. The *delay interval* (excess delay spread) is the length of time between the two farthest separated points of where the impulse response drops to given power level. The *correlation bandwidth* is defined as the bandwidth over which a given level of correlation of the transfer function is met. The *averaged power delay profile* can be thought of as the square of the spatially averaged (over a few wavelengths) impulse response for the channel for a given time period. In this text, only the maximum and the rms delay spread are used.

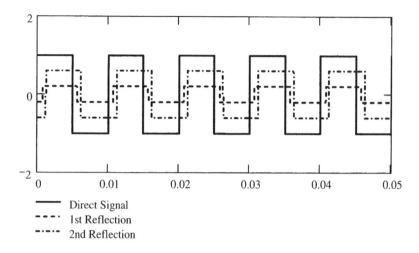

Direct Signal
--- 1st Reflection
--·-- 2nd Reflection

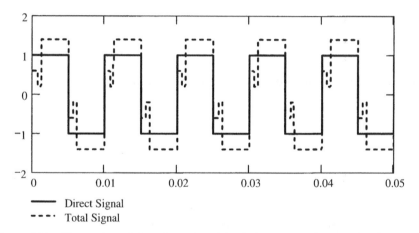

Direct Signal
--- Total Signal

Figure 8.22 Illustration of the effect of multipath delay spread on received symbols.

It is desirable to have the maximum delay spread to be small relative to the symbol interval of a digital communication signal. An analogous requirement is that the coherence bandwidth be greater than the signal bandwidth. Coherence bandwidth is defined as the bandwidth over which the channel can be considered flat with linear phase. Flat frequency response with linear phase implies no signal distortion. The coherence bandwidth is often approximated as

$$B_c \approx \frac{1}{5\sigma_t} \quad \text{to} \quad \frac{1}{50\sigma_t} \tag{8.24}$$

depending upon the flatness required and the channel's spectral shape.

If the signal bandwidth is less than the coherence bandwidth, $B < B_c$, then the channel is considered wideband or flat (flat fading). Otherwise, it is called a narrowband channel (selective fading). Flat fading causes the amplitude of the received signal to vary, but the spectrum of the signal remains in tact. The remedy for flat fading is to determine the allowable outage time and then use the Rayleigh pdf to determine the required fade margin to meet the requirement. For selective fading, some modulations are relatively tolerant of frequency dropouts, whereas in other cases an equalizer may be used.

Example 8.6. Given a digital communication system with a symbol rate of 50,000 symbols per second, what is an acceptable amount of rms delay spread?

For digital communication systems, the signal bandwidth is often approximated by the symbol rate,

$$B \approx R_S$$

So,

$$B \approx 50\,\text{kHz}$$

which is then set equal to the coherence bandwidth

$$B_c = 1/(5\sigma_t)$$

which implies

$$\sigma_t = 1/(250\,\text{kHz}) = 4\,\text{ms}$$

So the rms delay spread should ideally be less than or equal to 4 ms.

Another approach is simply consider the symbol time,

$$R_s = 50,000\,\text{baud} \rightarrow T \approx 20\,\text{ms}$$

Then to have a delay spread that is significantly less than T, $\sigma_t \ll T$, a value of about one-tenth of the symbol period can be used:

$$\sigma_t = 2\,\text{ms}$$

Thus a value of rms delay spread of 2–4 ms should be acceptable. □

Example 8.7. Given the communication system and the environment shown in Figure 8.23, if all of the buildings are good reflectors and both the transmitter and receiver are using wide-angle antennas, what is the maximum expected delay spread value? Base on that value, what is the highest symbol

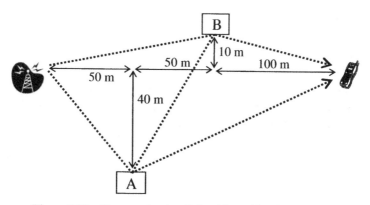

Figure 8.23 Communication link with multipath reflectors.

rate that you would recommend? Assume that there is no equalizer present and that both ends of the link are at the same height.

The direct path is 200 m and the longest reflected path bounces off of both buildings (i.e., Tx to A to B to Rx). The multireflection path is

$$d_{mr} = \sqrt{2500 + 1600} + \sqrt{2500 + 2500} + \sqrt{100 + 10000} \text{ m}$$

or

$$d_{mr} = 235.2 \text{ m}$$

Since the direct path is 200 m, the excess path length for the worst-case multireflection path is 35.2 m, which corresponds to a worst-case delay spread value of 117 ns. If this is equal to 10% or less of a symbol duration, then the multipath delay spread will be tolerable. Thus the minimum symbol time is 1.17 μs, which corresponds to a maximum symbol rate of 852 ksps. Note that for this problem the maximum delay spread rather than the rms delay spread was used since the specific geometry of the reflectors was well-defined. □

It is interesting to contrast the effect of delay spread with the signal fades that occur due to Fresnel zone boundary interference. The reflection points from the Fresnel zone computation resulted in time shifts on the order of the period of the carrier. The reflection points of concern for delay spread are generally much farther removed since they result in time shifts on the order of the symbol interval. When delay spread is a factor, the effects of carrier cancellation are incorporated into the multipath fade statistics (i.e., averaged in) and are not treated separately. Stated another way, if the magnitude of the multipath delay is sufficient to cause delay spread on the signal, then it is too large to permit meaningful characterization of carrier cancellation effects, and these effects are treated statistically.

8.4.2 Doppler Spread

Relative motion between the transmitter and receiver imparts a Doppler shift on the signal, where the entire signal spectrum is shifted in frequency. When multipath is combined with relative motion, the electromagnetic wave may may experience both positive and negative Doppler shift, smearing or spreading the signal in frequency. This effect is called *Doppler spread*. Figure 8.24 shows how this spreading could occur in an urban mobile telecommunications environment. In this figure, as the car moves to the right, the reflections from in front of the vehicle will have a positive Doppler shift and the signal from the tower will have negative Doppler shift. The magnitude of the Doppler shifts depends upon the geometry of the signal path (or reflection as shown). A related parameter, called *coherence time*, is defined as the time duration over which the signal amplitude will be highly correlated (close to constant):

$$T_c \cong \frac{1}{f_m} \tag{8.25}$$

where f_m is the maximum Doppler shift that is encountered:

$$f_m = \frac{\Delta v_{Rel}}{\lambda} \tag{8.26}$$

and Δv_{Rel} is the maximum relative velocity.

The coherence time is sometimes taken to be

$$T_c \cong \frac{0.423}{f_m} \tag{8.27}$$

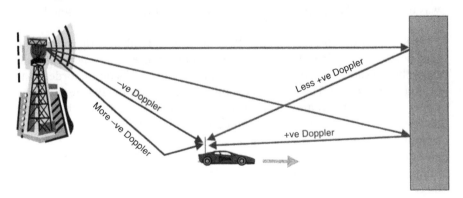

Figure 8.24 Illustration of how doppler spreading can occur.

If the signal bandwidth is much larger than twice the maximum Doppler shift, the channel is called a *slow-fading channel* and the effects of Doppler spread are negligible. A similar requirement is

$$T_{Sym} < T_c$$

(i.e., the symbol time is less than the coherence time of the channel). Thus the speed characterization of small-scale fading (fast or slow) depends upon the relationship between the Doppler bandwidth and the signal bandwidth, or equivalently the relationship between the coherence time and the symbol time. For a fast-fading channel, the receiver must be able to tolerate fades within a symbol, which makes amplitude-dependent modulations such as QAM a poor choice. For a slow-fading channel, however, the channel can be assumed constant over a symbol interval so amplitude modulations can be used. Note that a slow-fading channel may still experience very deep fades, which may last for one or more symbol durations. This can be addressed by using extra link margin and fast AGC, by increasing the modem interleaver depth to span the fades, by using short frame lengths and data retransmission, or by other methods.

8.4.3 Channel Modeling

The impulse response of the channel may be modeled as a time-varying impulse response $h(t_1, t)$. If the channel is time-varying and linear and has multipath present, it can be modeled by a delay-Doppler-spread function, also called a *scattering function*. The scattering function consists of the sum of delayed, attenuated, and Doppler-shifted versions of the signal [19].

In a slow-fading channel the amplitude of the signal remains relatively constant over a symbol interval, and therefore the channel response can be treated as being time-invariant over the symbol. With this restriction, t_1 is fixed and a time-invariant impulse response can be used. The received signal is then computed as the transmitted signal convolved with the channel impulse response:

$$r(t) = s(t) * h(t)$$

The impulse response must be causal (no output before the input is applied) and thus $h(t)$ is equal to zero for $t < 0$. Automatic gain control (AGC) at the receiver can be used to stabilize the receive signal level by adjusting the receiver gain at a rate slower than the symbol rate yet faster than the fade rate. If the channel is frequency-selective, this impulse response is used to model the channel, and it is then possible to synthesize an equalizing filter to eliminate the multipath distortion. Since channels generally vary over time, adaptive equalizers are used, which change in response to changes in the channel. Some equalizers use a special training sequence embedded within the signal, and others adapt based on the signal characteristics alone.

In a fast-fading channel, the signal amplitude will change during a symbol. In this case, the detector must be designed to accommodate the rapid amplitude variations. If a rapid-response AGC was used, the effect would be to track out any amplitude modulation component of the signal as well. Such channels are good candidates for phase modulation, but amplitude or phase/amplitude (QAM) modulations will not perform well. If the fast-fading channel is frequency-selective, the receiver must be able to tolerate the frequency distortion since equalization cannot generally be performed fast enough.

Orthogonal frequency division multiplexing (OFDM) can be used to combat fast fading and frequency-selective fading. OFDM uses multiple carriers, each operating at a reduced symbol rate to mitigate the effects of delay spread and selective fading. The longer symbol times make fast fading less of an issue, and redundancy over the carrier frequency reduces the effect of any frequency dropouts.

Table 8.1 shows the relationships between the different classes of small-scale fading and provides some comments on mitigation of the effects and limitations on the modulation that can be used.

8.4.4   The Probabilistic Nature of Small-Scale Fading

The probability of various fade depths depend upon the characteristics of the multipath environment. When there are a sufficient number of reflections present and they are adequately randomized (in both phase and amplitude), it is possible to use probability theory to analyze the underlying probability density functions (pdf's) and thereby determine the probability that a given fade depth will be exceeded. For the non-LOS case, where all elements of the received signal are reflections or diffraction components and no single component is dominant, the analysis draws on the central limit theorem to argue that the in-phase and quadrature components of the received signal will be independent, zero-mean Gaussian random variables with equal variance. In this case, the pdf of the received signal envelope will be a Rayleigh random variable. Using the Rayleigh pdf, the analyst can then determine the probability of any given fade depth being exceeded. It is also possible to optimize the system detection and coding algorithms for Rayleigh fading rather than for AWGN.

Example 8.8. Given a non-LOS communication link that experiences Rayleigh fading, what is the probability of a 12-dB or greater fade?

When considering fade depth, the fade is measured relative to the mean signal value, so the parameters for the Rayleigh are all relative and the absolute values are not required. The expression for the probability of a fade of x dB is obtained by integrating the Rayleigh pdf (Appendix A) from zero amplitude (the minimum value of the Rayleigh random variable) to the fade point:

TABLE 8.1 Relationship Between Different Small-Scale Fading Characterizations

	Delay Spread Effects	Comments		Doppler Effects	Comments
Slow Fading	$T_c > T_s$ Correlation time is greater than a symbol interval.	AGC is helpful, amplitude information can be preserved. Can use PSK or QAM.	Flat	$B_c > B_s$ Correlation bandwidth exceeds the signal bandwidth.	No need to equalize.
			Frequency-Selective	$B_c < B_s$ Correlation bandwidth is less than the signal bandwidth.	Can equalize.
Fast Fading	$T_c < T_s$ Correlation time is less than a symbol interval.	ACG cannot eliminate amplitude variations, amplitude information is lost. Can use PSK, but not QAM.	Flat	$B_c > B_s$ Correlation bandwidth exceeds the signal bandwidth.	No need to equalize.
			Frequency-Selective	$B_c < B_s$ Correlation bandwidth is less than the signal bandwidth.	Cannot usually equalize.

$$P_x = \int_0^{\sigma 10^{-x/20}} \frac{r}{\sigma^2} e^{\frac{-r^2}{2\sigma^2}} dr$$

where the fade point is referenced to the average signal power, σ^2. This integral can be solved to yield a closed form.

$$P_x = -e^{\frac{-r^2}{2\sigma^2}} \Big|_0^{\sigma 10^{-x/20}}$$

$$P_x = 1 - e^{-\frac{1}{2}10^{-x/10}}$$

Using this expression, the probability of a 12-dB fade is found to be 0.031, which means that the signal will be more than 12 dB below the mean signal level about 3.1% of the time. □

Figure 8.25 shows plots of probability versus fade depth. The left-hand plot includes negative fades (signal enhancement), while the right-hand plot has an expanded fade depth scale.

On the other hand, if there is an LOS component present or if one of the reflections is dominant, then the envelope of the received signal is characterized by a Ricean pdf, rather than a Rayleigh. The determination of the probability of a fade is a little more complicated when the Ricean pdf is used, but it is still tractable.

Example 8.9. Given a non-LOS communication link that has a dominant path in addition to other weaker reflective paths, what is the probability of a 12-dB or greater fade? The average power in the dominant path is equal to twice the average received power.

The dominant reflection implies that the fading will be governed by a Ricean pdf. The required parameters for the Ricean pdf (Appendix A) are as follows:

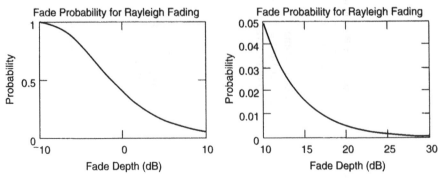

Figure 8.25   Probability of fade versus fade depth for a Rayleigh fading signal.

σ^2 is the *average power* from all of the reflections (i.e., the total average received power)

A is the *amplitude* of the dominant reflection

K is the Ricean factor ($K = 10\log[A^2/(2\sigma^2)]$)

Let the average received multipath power be represented by σ^2. The average power in the dominant (or specular) path is equal to $A^2/2$. Then for this example,

$$A^2/2 = 2\sigma^2$$

Thus

$$A = 2\sigma$$

$$P_x = \int_0^{\sigma 10^{-x/20}} \frac{r}{\sigma^2} e^{\frac{-(r^2+A^2)}{2\sigma^2}} I_0\left(\frac{Ar}{\sigma^2}\right) dr$$

Substituting for A, yields the desired expression

$$P_x = \int_0^{\sigma 10^{-x/20}} \frac{r}{\sigma^2} e^{\frac{-(r^2+4\sigma^2)}{2\sigma^2}} I_0\left(\frac{2r}{\sigma}\right) dr$$

While not trivial, this integral can be evaluated using a computational aid such as Mathcad to obtain the probability of a 12-dB or greater fade of 0.018.

Figure 8.26 shows several curves of the probability of exceeding a fade depth for several values of Ricean factor.

8.5 SUMMARY

For terrestrial communications, fading is roughly grouped into two types of fading: large-scale and small-scale. Large-scale fading encompasses those effects that change the path loss when the relative geometry of the transmitter and receiver has changed by several wavelengths or more. Such effects include increased free-space loss, ground-bounce loss, and shadowing. Small-scale fading is fading that occurs due to movement on the order of a wavelength and is primarily attributed to multipath. The expression for ground-bounce multipath indicates that the path loss will be proportional to the fourth power of the link distance and will be independent of the wavelength when communication takes place over a smooth reflective surface. The Rayleigh criterion is used to determine whether the surface can be charactgerized as smooth or not.

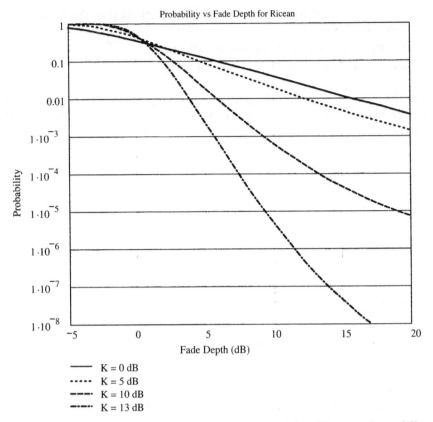

Figure 8.26 Probability of exceeding a given fade depth for different values of K.

Fresnel zone analysis is used to determine if any objects in the field of view may be affecting propagation. By keeping 60% of the first Fresnel zone clear of obstructions, the effect of stray reflection and diffraction is minimized. For a large obstacle in the propagation path, it is possible to model the effect of diffraction over the obstacle and estimate the wave strength at the receiver. An obstacle that is below the line of sight may still diffract the electromagnetic wave and cause the received wave to be attenuated. When there is no single or dominant obstacle in the proagation path, but the receiver is shadowed by a large number of random objects, the receiver is picking up a signal that has been attenuated (diffracted) by many different elements. The received signal is affected by each of these elements in a multiplicative manner, and when considered in dB the multiplication becomes addition. If the attenuations are independent, then this becomes the sum of many independent random variables, which tends to a Gaussian pdf as the number of contributions tends to infinity. Thus when multiple obstacles or diffractors are present, log-normal fading is used to model the path loss or received signal level.

Once the large-scale fading has been characterized, the analyst may then turn his or her attention to the small-scale fading effects, which occur in reponse to changes in the tramsitter to receiver geometry on the order of a wavelength. Small-scale fading is characterized as either fast- or slow-fading. Slow fading is attributed to the Doppler spread between the direct signal and the reflected (multipath signals). Slow fading is defined as the correlation time of the fade being greater than the symbol interval so that the fade may be treated as constant over the symbol. Receiver AGC can be used to adjust the receiver to keep the amplitude of successive symbols consistent. For fast fading, AGC is not helpful and the use of amplitude-dependent modulations is not recommended.

Small-scale fading is also characterized by its spectral properties, either flat or frequency selective. The spectral properties of the channel are dictated by the delay spread, the variation in delay between the direct and reflected signals. Excessive delay spread can lead to intersymbol interference. If the channel is slow-fading and frequency-selective, then an equalizer can be used to compensate for the spectral effects of the channel.

REFERENCES

1. L. V. Blake, *Radar Range Performance Analysis*, Munro Publishing Co., Silver Spring, MD, 1991, pp. 261–263.
2. J. D. Parsons, *The Mobile Radio Propagation Channel*, 2nd ed., Wiley, West Sussex, 1992, pp. 21–22.
3. W. C. Y. Lee, *Mobile Communication Engineering, Theory and Applications*, 2nd ed., McGraw-Hill, New York, 1998, pp. 109–114.
4. S. R. Saunders, *Antennas and Propagation for Wireless Communication Systems*, Wiley, West Sussex, 1999, pp. 46–51.
5. N. Blaunstein, *Radio Propagation in Cellular Networks*, Artech House, Norwood, MA, 2000, pp. 48–53.
6. J. D. Parsons, *The Mobile Radio Propagation Channel*, 2nd ed., Wiley, West Sussex, 1992, pp. 33–34.
7. W. L. Stutzman, G. A. Thiele, *Antenna Theory and Design*, 2nd ed., Wiley, Hoboken, NJ, 1998, p. 556.
8. T. S. Rappaport, *Wireless Communications, Principles and Practice*, 2nd ed., Prentice-Hall, Upper Saddle River, NJ, 2002, pp. 129–134.
9. W. C. Y. Lee, *Mobile Communication Engineering, Theory and Applications*, 2nd ed., McGraw-Hill, New York, 1998, p. 142.
10. ITU-R Recommendations, *Propagation by Diffraction*, ITU-R P.526-7, Geneva, 2001.
11. T. S. Rappaport, *Wireless Communications, Principles and Practice*, 2nd ed., Prentice-Hall, Upper Saddle River, NJ, 2002, pp. 134–136.
12. J. D. Parsons, *The Mobile Radio Propagation Channel*, 2nd ed., Wiley, West Sussex, 1992, pp. 45–46.

13. B. McLarnon, VHF/UHF/Microwave Radio Propagation: A Primer for Digital Experimenters, www.tapr.org
14. W. C. Y. Lee, *Mobile Communication Engineering, Theory and Applications*, 2nd ed., McGraw-Hill, New York, 1998, pp. 147–149.
15. N. Blaunstein, *Radio Propagation in Cellular Networks*, Artech House, Norwood, MA, 2000, pp. 135–137.
16. J. D. Parsons, *The Mobile Radio Propagation Channel*, 2nd ed., Wiley, West Sussex, 1992, pp. 42–45.
17. S. R. Saunders, *Antennas and Propagation for Wireless Communication Systems*, Wiley, West Sussex, 1999, Chapter 9.
18. S. R. Saunders, *Antennas and Propagation for Wireless Communication Systems*, Wiley, West Sussex, 1999, pp. 186–188.
19. ITU-R Recommendations, *Multipath propagation and parameterization of its characteristics*, ITU-R P.1407-0, Geneva, 1999.

EXERCISES

1. Given a communication link with each transceiver mounted on a 20-m tower, operating over a smooth, reflective terrain, what is the path loss experienced over the path if the transceivers are 2 km apart? Assume the frequency is 10 GHz and the antennas each have unity gain (0 dB). Justify your answer.

2. Consider a 1.9-GHz system with the following parameters:

$$h_m = 5 \text{ m}, \qquad h_t = 30 \text{ m}$$
$$G_t = 20 \text{ dBd}, \qquad G_m = 0 \text{ dBd}$$

If the maximum allowable path loss for reception is 148 dB, what is the maximum distance at which 90% coverage can be obtained? You may use the Lee Model for the median path loss with $\gamma = 43.1$ and $L_0 = 106.3$ dB. Assume log-normal fading with $\sigma_t = 8$ dB and use $n = 2.5$ for the frequency adjustment exponent in F_5.

3. Consider a PCS cell that is located in a relatively rural area, where there is only a small amount of blockage or shadowing, and free-space loss can be used for the median path-loss computation. Given the system parameters

$$f = 1.9 \text{ GHz}$$
$$\sigma_S = 3 \text{ dB (location variability)}$$
$$PL_{max} = 110\text{-dB maximum allowable path loss}$$

what is the maximum distance at which 95% coverage can be expected? Assume that the shadowing is log-normal.

4. A millimeter-wave (28-GHz) data link uses a 0.3-m-diameter aperture antenna on each end and a 20-dBm transmitter. The waveguide loss at the back of the antenna is 1 dB and the radome loss is 0.5 dB. If the signal bandwidth is 10 MHz, the equivalent noise temperature is 800 K and the required CNR for detection is 25 dB, what is the maximum distance between the radios for reliable operation? Assume that the antenna efficiency is 60%, both radios are mounted 10 m above the ground, and the ground is very reflective. Also assume that a 20-dB fade margin is required. Generate a link budget and then compute the maximum distance.

5. A 1.9-GHz data link uses a 0.3-m diameter aperture antenna on each end and a 10-dBm transmitter. The waveguide loss at the back of the antenna is 1 dB and the radome loss is 0.5 dB. If the signal bandwidth is 1 MHz, the equivalent noise temperature is 800 K and the required CNR for detection is 15 dB, what is the maximum distance between the radios for reliable operation? Assume that the antenna efficiency is 60%, both radios are mounted 2 m above the ground, and the ground is very reflective. Also assume that a 20-dB fade margin is required. Generate a link budget and estimate the maximum range.

6. Consider the 2-km communication link shown in the accompanying diagram. If the frequency of the link is 2.85 GHz, the middle building is located between the other two as shown, and the roof is reflective, determine the following:

   • Can FSL be used to plan the link (i.e., is the middle building producing any significant blockage of the signal)? Justify your answer.
   • Will reflected signal from the roof cause a multipath type of signal fade (i.e., cancellation)? Again explain your conclusions.

   Assume that the antenna patterns are fairly broad, so that antenna discrimination does not factor into the analysis.

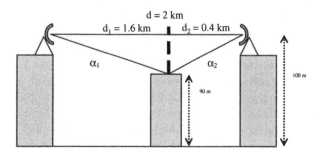

Diagram for Exercise 5.

7. Given a channel with a coherence bandwidth, B_c, of 100 KHz, is there likely to be any significant distortion or intersymbol interference to a wireless LAN operating at a symbol rate of 1 million symbols per second?

CHAPTER 9

Indoor Propagation Modeling

9.1 INTRODUCTION

Indoor propagation of electromagnetic waves is central to the operation of wireless LANs, cordless phones, and any other indoor systems that rely on RF communications. The indoor environment is considerably different from the typical outdoor environment and in many ways is more hostile. Modeling indoor propagation is complicated by the large variability in building layout and construction materials. In addition, the environment can change radically by the simple movement of people, closing of doors, and so on. For these reasons, deterministic models are not often used. In this chapter, some statistical (site-general) models are presented. When fit to empirical data, these models can provide a reasonable representation of the indoor environment.

9.2 INTERFERENCE

While an understanding of indoor propagation is essential, another important element of indoor wireless operation that should be considered is interference. Unlike outdoor environments, where the operating distances are greater, in an indoor environment, it is possible, and in fact common, to have an interfering system operating within a few feet or less of a given system. A classic but by no means isolated example is the desktop computer with a wireless LAN card that also employs a wireless keyboard and/or mouse. The wireless keyboard and mouse are likely to use the Bluetooth standard, which uses frequency hopping in the 2.4-GHz ISM band. If the wireless LAN card is an 802.11b or g [direct sequence spread spectrum (DSSS) or orthogonal frequency division multiplexing (OFDM)] system, then it will be operating in the same frequency band and the potential for interference exists. In addition, a computer has a variety of internal high-frequency digital clocks that generate harmonics, which may fall within the system's passband. A monitor may also produce a substantial amount of interference. Add to that the RF energy radi-

Introduction to RF Propagation, by John S. Seybold
Copyright © 2005 by John Wiley & Sons, Inc.

ated by fluorescent lighting, other consumer products, and office equipment, and it becomes clear that the interference environment indoors is very unfriendly. It is important to account for this interference and to understand that communication link problems in indoor environments may not be propagation issues, but rather interference issues. Sometimes simply repositioning a piece of equipment by a few feet is enough to resolve the problem.

The effect of various interferers can be estimated by computing the signal-to-interference ratio at the receiver of interest or *victim receiver*. In general the signal-to-interference ratio should be at least as large as the required signal-to-noise ratio for acceptable operation. Another effect, called de-sensing, can occur even when the interferer is operating at a different frequency or on a different channel from the victim receiver. This can occur in different ways. The most straightforward effect is when the interfering signal is within the front-end bandwidth of the victim receiver and causes the victim receiver's AGC to reduce the front-end gain. Another effect is that the interfering signal may actually overload the front end of the victim receiver, causing nonlinear operation. This nonlinear operation will cause intermodulation products to be produced [1], some of which may occur within the operating channel, again resulting in reduced receiver sensitivity.

9.3 THE INDOOR ENVIRONMENT

The principal characteristics of an indoor RF propagation environment that distinguish it from an outdoor environment are that the multipath is usually severe, a line-of-sight path may not exist, and the characteristics of the environment can change drastically over a very short time or distance. The ranges involved tend to be rather short, on the order of 100 m or less. Walls, doors, furniture, and people can cause significant signal loss. Indoor path loss can change dramatically with either time or position, because of the amount of multipath present and the movement of people, equipment, and/or doors [2]. As discussed in the previous chapter, multiple reflections can produce signal smearing (delay spread) and may cause partial signal cancellation (signal fading).

9.3.1 Indoor Propagation Effects

When considering an indoor propagation channel, it is apparent that in many cases there is no direct line of sight between the transmitter and receiver. In such cases, propagation depends upon reflection, diffraction, penetration, and, to a lesser extent, scattering. In addition to fading, these effects, individually and in concert, can degrade a signal. Delay and Doppler spread are usually far less significant in an indoor environment because of the much smaller distances and lower speeds of portable transceiver as compared to outdoor environments. However, this advantage is usually offset by the fact that many indoor applications are wideband, having short symbol times and therefore a corresponding greater sensitivity to delay spread. In addition, the wave may experience depolarization, which will result in polarization loss at the receiver.

9.3.2 Indoor Propagation Modeling

There are two general types of propagation modeling: site-specific and site-general. Site-specific modeling requires detailed information on building layout, furniture, and transceiver locations. It is performed using ray-tracing methods in a CAD program. For large-scale static environments, this approach may be viable. For most environments however, the knowledge of the building layout and materials is limited and the environment itself can change, by simply moving furniture or doors. Thus the site-specific technique is not commonly employed. Site-general models [3–6] provide gross statistical predictions of path loss for link design and are useful tools for performing initial design and layout of indoor wireless systems. Two popular models, the ITU and the log-distance path loss models, are discussed in the following sections.

9.3.3 The ITU Indoor Path Loss Model

The ITU model for site-general indoor propagation path loss prediction [3] is

$$L_{total} = 20\log(f) + N\log(d) + Lf(n) - 28 \text{ dB} \tag{9.1}$$

where

N is the distance power loss coefficient
f is the frequency in MHz
d is the distance in meters ($d > 1\,\text{m}$)
$Lf(n)$ is the floor penetration loss factor
n is the number of floors between the transmitter and the receiver

Table 9.1 shows representative values for the power loss coefficient, N, as given by the ITU in Ref. 3, and Table 9.2 gives values for the floor penetration loss factor.

The ITU model can be shown to be equivalent to the equation for free-space loss with the distance power being $N = 20$ (when not traversing floors). Thus the ITU model is essentially a modified power law model. This can be seen as follows: The expression for free-space loss expressed in dB is given by

$$L_{dB} = -20\log(\lambda) + 20\log(4\pi) + N\log(d) \quad \text{dB}$$

when $N = 20$. The first term on the right-hand side can be expressed as

TABLE 9.1   Power Loss Coefficient Values, N, for the ITU Site-General Indoor Propagation Model

Frequency	Residential	Office	Commercial
900 MHz	—	33	20
1.2–1.3 GHz	—	32	22
1.8–2 GHz	28	30	22
4 GHz	—	28	22
5.2 GHz	—	31	—
60 GHz[a]	—	22	17

[a] 60 GHz is assumed to be in the same room.

Source: Table 2 from Ref. 3, courtesy of ITU.

TABLE 9.2   Floor Penetration Loss Factor, $Lf(n)$, for the ITU Site-General Indoor Propagation Model

Frequency	Residential	Office	Commercial
900 MHz	—	9 ($n = 1$) 19 ($n = 2$) 24 ($n = 3$)	—
1.8–2 GHz	$4n$	$15 + 4(n - 1)$	$6 + 3(n - 1)$
5.2 GHz	—	16 ($n = 1$ only)	—

n is the number of floors penetrated ($n \geq 1$).

Source: Table 3 from Ref. 3, courtesy of ITU.

$$20\log(\lambda) = 20\log(c) - 20\log(f(\mathrm{Hz}))$$

$$20\log(\lambda) = 169.54 - 20\log(f(\mathrm{MHz})) - 120 = 49.54 - 20\log(f(\mathrm{MHz}))$$

Using the fact that

$$20\log(4\pi) = 22$$

the expression for the path loss simplifies to

$$L_{dB} = 20\log(f) + N \cdot \log(d) - 27.54 \qquad \mathrm{dB}$$

which, when expressed with two significant digits, agrees with the ITU site-general indoor propagation model.

A few comments about the values in Table 9.1 and application of the ITU site-general model are in order. A power loss coefficient value of $N = 20$ corresponds to free-space loss, and this will usually apply in open areas. Corridors may channel RF energy, resulting in a power loss coefficient of $N = 18$

(slightly less than free-space loss). In the case of propagation around corners or through walls, $N = 40$ is used. For long paths, the reflected path(s) may interfere, resulting in $N = 40$ being used here as well.

Example 9.1. Consider application of a 5.2-GHz wireless LAN in an office building. If the longest link is 100 m, what is the maximum path loss? How much additional path loss will exist between floors? Is the interfloor loss sufficient to permit frequency re-use?

From the power loss coefficient values given in Table 9.1, for an office building the value of N is found to be

$$N = 31$$

The floor penetration loss factor for a single floor from Table 9.2 is

$$Lf = 16 \, dB$$

For $d = 100$ m, the expected path loss is then given by

$$L_{total} = 20 \log(f) + N \log(d) + Lf - 28 \, dB$$

where

$f = 5200 \, MHz$
$N = 31$
$d = 100$
$Lf = 16$

Thus the expression for the maximum expected total path loss is

$$L_{total} = 20 \log(5200) + 31 \log(100) - 12$$

so

$$L_{total} = 12 \, dB \text{ between two floors}$$

and

$$L_{total} = 108 \, dB \text{ on the same floor}$$

The 16-dB difference when penetrating a floor would probably not be sufficient to permit frequency re-use between adjacent floors without a significant risk of interference. This analysis assumed that the adjacent floor receiver would also be 100 m from the transmitter. In practice, the receiver on the adjacent floor may be closer than 100 m, it could be situated just above or below the transmitter. □

The ITU site-general indoor propagation model includes provisions for modeling the expected delay spread. As discussed in Chapter 8, delay spread represents a time-varying, smearing effect (in time) on the desired signal and can be characterized by a power delay profile. Based on the ITU model, the impulse response of the channel may be modeled as

$$h(t) = e^{-t/S} \qquad \text{for } 0 < t < t_{max}$$

$$h(t) = 0 \qquad \text{otherwise}$$

where

S is the rms delay spread

t_{max} is the maximum delay (obviously $t_{max} \gg S$)

A plot of this impulse response function is shown in Figure 9.1. Table 9.3 shows some typical values of rms delay spread at 2.4 and 5.2 GHz [3].

The rms delay spread characteristics of the indoor environment need to be considered by the communication system designer, and there is little that the

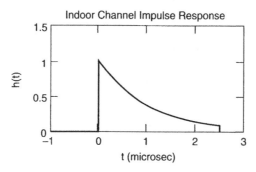

Figure 9.1   Representative indoor channel impulse response using the ITU model ($S = 1\,\mu s$, $t_{max} = 2.5\,\mu s$).

TABLE 9.3   Typical RMS Delay Spread Parameters at 1.9 and 5.2 GHz

Frequency	Environment	Often	Median	Rarely
1.9 GHz	Residential	20 ns	70 ns	150 ns
1.9 GHz	Office	35 ns	100 ns	460 ns
1.9 GHz	Commercial	55 ns	150 ns	500 ns
5.2 GHz	Office	45 ns	75 ns	150 ns

Source: Table 5 from Ref. 3, courtesy of ITU.

network designer can do beyond simply repositioning transmitters and receivers to avoid any unacceptable delay spread.

9.3.4 The Log-Distance Path Loss Model

The log-distance path loss model is another site-general model [4] and it is given by

$$L_{total} = PL(d_0) + N \cdot \log(d/d_0) + X_s \quad \text{dB} \qquad (9.2)$$

where

$PL(d_0)$ is the path loss at the reference distance, usually taken as (theoretical) free-space loss at 1 m

$N/10$ is the path loss distance exponent

X_s is a Gaussian random variable with zero mean and standard deviation of σ dB

The log-distance path loss model is a modified power law with a log-normal variability, similar to log-normal shadowing. Some "typical" values from Ref. 4 are given in Table 9.4.

Example 9.2. Consider an example similar to one used before before, the application of a 1.5-GHz wireless LAN in an office building. If the longest hop is 100 m, what is the maximum path loss if 95% coverage is desired?

The table for the log-distance path loss model indicates

$$N = 30$$

TABLE 9.4 Typical Log-Distance Path Model Parameter Measurements

Building	Frequency (MHz)	N	σ (dB)
Retail stores	914	22	8.7
Grocery store	914	18	5.2
Office, hard partition	1500	30	7.0
Office, soft partition	900	24	9.6
Office, soft partition	1900	26	14.1
Textile/Chemical	1300	20	3.0
Textile/Chemical	4000	21	7.0/9.7
Paper/Cereals	1300	18	6.0
Metalworking	1300	16/33	5.8/6.8

[a] Multiple entries are presumably from measurements at different times or facilities.

Source: Table 4.6 from Ref. 4, courtesy of Prentice-Hall.

The hard partition office building data at 1.5 GHz indicate

$$\sigma = 7.0$$

From the Q function table (Table A.1), $p = 0.05$ occurs when $z \cong 1.645$:

$$z = X_S/\sigma$$

so, the fade depth X_s is 11.5 dB.
    The predicted path loss is

$$L_{total} = PL(d_0) + N \cdot \log(d/d_0) + X_s, \qquad \text{where } d_0 = 1 \text{ m}$$
$$PL(d_0) = -20\log(\lambda) + 20\log(4\pi) + 20\log(1) = 36 \text{ dB}$$
$$N\log(d/d_0) = 30\log(100) = 60 \text{ dB}$$

And finally,

$$L_{total} = 36 + 60 + 11.5 = 107.5 \text{ dB}$$

For comparison, the free-space loss at this frequency and distance is 76 dB. □

Other indoor propagation models include (a) the Ericsson multiple break-point model [4], which provides an empirical worst-case attenuation versus distance curve, and (b) the attenuation factor model [4], which is similar in form to the log-distance model. There are a variety of empirical models for indoor propagation loss, most of which are modified power law. Any model used should either be optimized for the particular application if possible, or otherwise, additional margin included in the design. Rappaport [4] indicates that most log-distance indoor path loss models have about a 13-dB standard deviation, which suggests that a ±26-dB (two-sigma) variation about the pre-dicted path loss would not be unusual.
    Empirical path loss prediction models should be used to determine a high-level design. Once the design is complete, field measurements or an actual deployment serve as the final determination of a model's applicability to a given building. A good static design must still include sufficient margin for the environmental dynamics. Depending upon the bandwidth and data rate, delay-spread effects may also need to be considered. As a rule, the delay spread effects are considered by the modem or equipment designer and not by the individual who lays out the network, since the mitigation is best handled at the equipment level.

9.4 SUMMARY

In many ways, an indoor propagation channel is more hostile than a typical outdoor channel. Lack of a line of sight, heavy attenuation, diffraction by objects in the propagation path, and multipath all contribute to impairing a system's ability to communicate over an RF channel. In addition, the close proximity of interference sources and rapid variations in the channel make definitive or deterministic channel characterization difficult, if not impossible. There are two general classes of indoor propagation models: site-specific and site-general. The site-general models tend to be the more widely used models since site-specific models require a fairly static environment and considerable detail about the building layout and construction.

Two popular site-general models are discussed in this chapter: the ITU model and a log-distance model presented by Rappaport. The ITU model is a modified power law that uses empirical building data to predict the path loss. The ITU model also provides a model for the impulse response of the indoor channel to account for delay spread, again using empirical data. The log-distance model is a combination of a modified power law and a log-normal fading model that also uses empirical data. As with all models like these, it is best to ultimately take field measurements to verify that the model is accurately characterizing the environment. If it is found to not be representative, the model can be fine-tuned using the collected data.

REFERENCES

1. S. C. Cripps, *RF Power Amplifiers for Wireless Communications*, Artech House, Norwood, MA, 1999, Chapter 7.
2. D. Dobkin, Indoor propagation issues for wireless LANs, *RF Design Magazine*, September 2002, pp. 40–46.
3. ITU-R Recommendations, *Propagation Data and Prediction Methods for the Planning of Indoor Radiocommunication Systems and Radio Local Area Networks in the Frequency range 900 MHz to 100 GHz*, ITU-R P.1238-2, Geneva, 2001.
4. T. S. Rappaport, *Wireless Communications Principles and Practice*, 2nd ed., Prentice-Hall, Upper Saddle River, NJ, 2002, pp. 161–166.
5. H. Hashemi, The indoor radio propagation channel, *Proceedings of the IEEE*, Vol. 81, No. 7, July 1993, pp. 943–968.
6. J. Kivinen, X. Zhoa, and P. Vainikainen, Empirical characterization of wideband indoor radio channel at 5.3 GHz, *IEEE Transactions on Antennas and Propagation*, Vol. 49, No. 8, August 2001, pp. 1192–1203.

EXERCISES

1. What is the median expected path loss at 100 m in an office building if the frequency is 1.9 GHz? Use the ITU indoor propagation model.

2. Generate a plot of the path loss for problem 1 versus the probability of that path loss occurring.

3. Repeat problem 1 using the log-distance model and assume that 98% coverage is required at the fringe (i.e., at $d = 100\,\text{m}$) and that the building contains soft partitions.

4. Consider an indoor communications link operating at 900 MHz in an office building with hard partitions. If the link distance is 38 m, what minimum and maximum path loss may be encountered? Use the log-distance path model and values at the 99th percentile (i.e., what is the minimum and maximum loss that will be encountered with a 99% probability).

5. Considering the rms delay spread parameters given in Table 9.3, what is the highest practical symbol rate that can be employed without incorporating equalization into the modem. Explain your reasoning.

CHAPTER 10

Rain Attenuation of Microwave and Millimeter-Wave Signals

10.1 INTRODUCTION

A key performance metric for most communication systems is the availability—that is, the percentage of time that the link is providing communications at or below the specified bit error rate. There are a number of factors that impact the availability, including hardware reliability, interference, and fading. As discussed in earlier chapters, there are a variety of sources of excess path attenuation, including atmospheric absorption, diffraction, multipath effects (including atmospheric scintillation), shadowing, foliage, and attenuation by hydrometeors. Of these, the fades caused by rain (hydrometeor) attenuation can be a major limiting factor of link availability or link distance. This is particularly true at millimeter-wave frequencies where the fade depths can be severe. Rain fades start to become a concern above 5 GHz and, by 20–30 GHz, can be a significant factor depending upon the link distance and the geographic location.

The amount link margin allocated to rain fades, the communication link distance, and the local climate all factor into determination of the rain availability. Rain availability is essentially the percentage of time that the available rain fade margin is not exceeded. It is important that the rain fade margin not be used for other margins unless the other factor is exclusive of rain. There is always the temptation to argue that the rain margin is not being used most of the time, so it can be used to compensate for other link budget shortfalls. The problem with this reasoning is that the rain fade margin may not be available when needed, resulting in rain availability below the intended value.

A considerable body of data and work exists for characterizing the statistical impact of rain [1–4]. There has also been work done on theoretical support for the empirical rain attenuation models [5–7]. Even so, the area of rain fade modeling is not considered a mature field. The statistics vary considerably by location, time of year and even from year-to-year. There are also a variety of

---

Introduction to RF Propagation, by John S. Seybold
Copyright © 2005 by John Wiley & Sons, Inc.

types of rain, which have different effects. Rain fades are very frequency-dependent, and results at one frequency must be carefully adjusted if they are applied at a different frequency [8, 9]. The primary motivation for scaling rain attenuation results over frequency is if a significant body of data exists for the geographical area of interest at a frequency other than the operating frequency and one wishes to scale the loss to the operating frequency. Knowledge of the physic involved forms the underpinnings of the frequency scaling. It is also possible to scale attenuation data from one polarization to another [8].

This chapter provides the details of two well-known models, the Crane global model and the ITU model, and their application to terrestrial links. The details of applying the models to satellite links and slant paths are presented in Chapter 11. Central to understanding the application of rain fade analysis is the concept of link availability and how it relates to the link budget. In general, a communications link designer must decide (or be told) what percentage of time the link must be operational. This availability is then allocated between the various sources that can cause link outages, including rain fades, interference, and hardware failures. Once a rain availability allocation is determined, the rain fade models can be applied to determine what level of fade will not be exceeded with probability equal to the rain availability allocation. That rain fade value is then incorporated into the link budget, and the resulting link budget can be used to determine either the maximum link distance or some other key parameter such as the required transmit power.

Once a model has been implemented, it is possible to parameterize the results on any of a number of key parameters for performing trade studies. Once such parameterization is to plot the total clear-air link margin versus link distance and then superimpose a curve of the rain fade depth versus link distance for a given availability, frequency, and polarization. Such a link distance chart provides a very concise assessment of the possible link performance. It is also possible to plot the availability versus link distance for a given link gain, frequency, and polarization. Link distance charts are discussed in Section 10.4.

It is sometimes of interest to determine the link availability when all of the other parameters are known. An example of this might be deploying a commercial system in a particular climate at a fixed distance. The operator would like to have a good estimate of the probability of a rain outage. This can be determined by using the availability function of the ITU model and is presented in detail in Section 10.5.

This chapter concludes with a brief discussion of the cross-polarization effect of rain and the effects of other types of precipitation.

10.2 LINK BUDGET

The principal limitation on millimeter-wave link availability is precipitation. While the hardware designer cannot account for rain, the link or network planner can and must, by incorporating sufficient margin into the link design.

Detailed rain attenuation models coupled with extensive rain statistics permit the link designer to trade link distance for availability for a given system.

The estimated maximum link distance is based on system gain, antenna gain, and propagation effects. The gains and the propagation effects must be known with high confidence for a statistical rain fade analysis to be meaningful, especially at high availabilities.

System gain, as used here, is defined as the maximum average transmit power minus the receiver sensitivity (in dB).

$$G_S = P_{Tmax} - R_{thresh} \quad \text{dB} \tag{10.1}$$

The *link gain* is the *system gain* plus the sum of the transmit and receive antenna gains in dB.

$$G_L = G_S + 2G_{Ant} = P_{Tmax} - 2G_{Ant} - R_{thresh} \quad \text{dB} \tag{10.2}$$

The so-called "typical" specification values for these parameters cannot be used in these calculations, and in fact even the three-sigma values for these parameters are not accurate enough if the desired availability is greater than about 99%. For instance, commercial antennas may have a published "typical" gain, but the actual antenna gain must be verified, because it will likely deviate from the typical value. Higher confidence in the parameter values is required and in fact it may be necessary to measure each individual unit. The designer may also want to de-rate the link gain by a dB or two for mutual interference and other factors. The *adjusted link gain* is defined as the *link gain* minus all other sources of fading and loss except rain.

Once the system has been characterized, the free-space loss equation can be applied:

$$L(d) = 20 \log(\lambda/(4\pi d)) \quad \text{dB}$$

The maximum theoretical, free-space link distance is achieved when the link gain is equal to the free-space loss, $L(d) = G_L$. For a given link distance, d, $G_L - L(d)$ gives the available fade margin, which can be allocated to overcome rain fades or other link impairments.

Figure 10.1 shows what a typical millimeter-wave communication system link budget might look like. The rain-fade entry is a fade margin that is computed using a rain model for the desired availability and geographical location and the link distance.

A plot of the predicted specific attenuation (loss in dB/km) due to atmospheric absorption versus frequency was provided in Figure 6.4 and is presented here in Figure 10.2. The total amount of absorption loss is determined by multiplying the specific attenuation at the operating frequency by the link distance.

Tx power	20 dBm				
Tx loss	−1.5 dB				
Tx antenna gain	36 dB				
Radome loss	−2 dB				
EIRP	**52.5 dBm**				
		freq =	38.6 GHz	lambda =	0.007772
Path loss (FSL)	−125 dB	d =	1.1 km	PL =	−125.001
Alignment error	−2 dB				
Rain fade	−51 dB				
Multipath	−2 dB				
Atmospheric loss	−0.13 dB			Gamma =	0.12 dB/km
Interference	−1 dB				
Total path losses	**−181.1 dB**				
Radome loss	−2 dB				
Rx antenna gain	36 dB			Link gain =	181.5
Rx loss	−1 dB				
Total Rx gain	**33 dB**			Adjusted link gain =	176.4
RSL	−95.6 dBm				
Sensitivity	−96.0 dBm				
Net margin	**0.4 dB**				

Figure 10.1 Typical link budget for a millimeter-wave communication link.

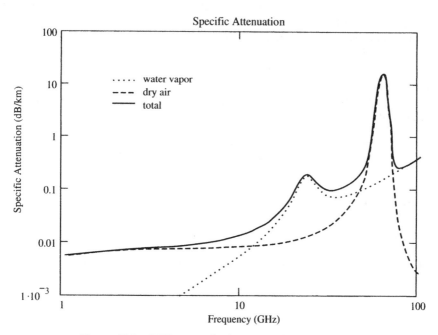

Figure 10.2 ITU atmospheric attenuation prediction.

10.3 RAIN FADES

Rain fades depend upon the rain rate, drop size and shape, and volume density (# drops per m³). Of these factors, only rain rate is readily measured unless a radar system is available; for this reason, rain rate is most often used for rain fade characterization. Robert Crane [2, 4] and the ITU [10–12] each provide rain fade models based on rain rate. Application of these models is presented in the following sections. There are other rain fade models available, but the models presented here are based on a significant amount of empirical data and are the most widely used.

10.3.1 Specific Attenuation Due to Rainfall

The ITU and Crane models both make use of the specific attenuation due to rain that is computed from the ITU data library. The specific attenuation is determined by using regression coefficients and the rain rate of interest. The models differ in the values for rainfall rate and in the modeling equations used, but they share the same regression coefficients for the specific attenuation. To determine the specific attenuation for a given rain rate, the frequency of operation is used to select the appropriate linear regression coefficients [10, 13] and interpolate them if necessary. Table 10.1 gives the linear-regression coefficients for linear polarization at several frequencies. Table 10A.1 in Appendix 10A is a more complete table of regression coefficients. The interpolation of the coefficients is performed using a log scale for frequency and the k values and a linear scale for the α values. Table 10A.2 provides interpolated regression coefficients for frequencies from 1 through 40 GHz in 1-GHz steps. Note that these coefficients are frequency- and polarization-dependent. The final coefficients are determined using the following expressions, which account for the path elevation angle and the polarization. Note that for circular polarization, a tilt angle, τ, of 45 degrees is used.

$$k = \frac{\left[k_H + k_V + (k_H - k_V)\cos^2(\theta)\cos(2\tau)\right]}{2} \tag{10.3}$$

TABLE 10.1   Regression Coefficients for Estimating Specific Attenuation

Frequency (GHz)	k_H	k_V	α_H	α_V
2	0.000650	0.000591	1.121	1.075
6	0.00175	0.00155	1.308	1.265
8	0.00454	0.00395	1.327	1.310
10	0.0101	0.00887	1.276	1.264
12	0.0188	0.0168	1.217	1.200
20	0.0751	0.0691	1.099	1.065
30	0.187	0.167	1.021	1.000
40	0.350	0.310	0.939	0.929

Source: Table 1 from Ref. 1, courtesy of the ITU.

$$\alpha = \frac{\left[k_H\alpha_H + k_V\alpha_V + (k_H\alpha_H - k_V\alpha_V)\cos^2(\theta)\cos(2\tau)\right]}{2k} \quad (10.4)$$

where

θ is the elevation path angle

τ is the polarization tilt angle (0, 45, and 90 degrees for horizontal, circular, and vertical, respectively)

The values in Table 10.1 are used by both the ITU and Crane models and apply equally to both terrestrial and satellite links when incorporated into the models. The coefficients suggest that the effect of rain on horizontally polarized signals is greater than that for vertically polarized signals. This is in fact true and is generally attributed to the vertically elongated shape of most raindrops. The effect on circularly (right-hand or left-hand) polarized signals is in between the two as might be expected. Figure 10.3 shows plots of the specific attenuation for a 50-mm/h rainfall rate at frequencies up to 40 GHz. The indicated data points are where the given ITU regression coefficients occur.

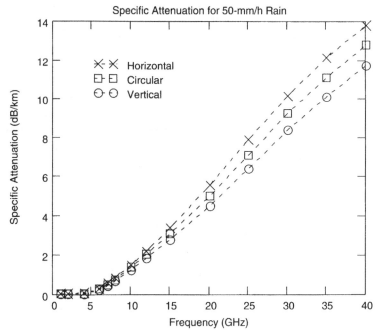

Figure 10.3 Comparison of specific attenuations for different polarizations versus frequency.

10.3.2 The ITU Model

The ITU rain attenuation model is given by Ref. 8. The first step in applying the ITU model for a given availability on a horizontal or nearly horizontal communications link is to determine the 99.99% fade depth.

$$\text{Atten}_{0.01} = k \cdot RR^{\alpha} \cdot d \cdot r \quad \text{dB} \tag{10.5}$$

where

> RR is the 99.99% rain rate for the rain region, in mm/h
> $k \cdot RR^{\alpha}$ is the specific attenuation in dB/km
> d is the link distance in km

and

$$r = 1/(1 + d/d_0) \tag{10.6}$$

with

$$d_0 = 35e^{-0.015 \cdot RR} \quad \text{km} \tag{10.7}$$

The parameter d_0 is the effective path length and r is called the *distance factor*. The specific attenuation is computed using the 99.99% rain rate for the desired location and season and the appropriate regression coefficients, k and α, for the frequency and polarization in use as given in Table 10A.1 and Table 10A.3 [10–14]. The rain rates based on geographical regions [11] are the most widely used and easily applied method of determining the rain rate. The ITU has released an updated model [12] that is discussed in Section 10.3.4.

These factors are used to model the fact that rain attenuation is not linear with distance. Simply multiplying the specific attenuation by the link distance does not provide an accurate estimate of the rain fade in most circumstances. Table 10.2 gives the 99.99% availability rain rates for the ITU model [11]. Data for other availabilities are also included in Appendix 10A. The procedure just described is validated for frequencies up to at least 40 GHz and distances up to 60 km [8].

The fade depths for availabilities other than 99.99% can be found using other data, or preferably by applying an adjustment factor [8].

TABLE 10.2 ITU Rain Rate Data for 0.01% Rain Fades

A	B	C	D	E	F	G	H	J	K	L	M	N	P
8	12	15	19	22	28	30	32	35	42	60	63	95	145

Source: Table 1 from Ref. 12, courtesy of the ITU.

$$\mathrm{Atten}/\mathrm{Atten}_{0.01} = 0.12p - (0.546 + 0.043\log(p)) \qquad (10.8)$$

for latitudes greater than 30 degrees, North or South, and

$$\mathrm{Atten}/\mathrm{Atten}_{0.01} = 0.07p - (0.855 + 0.139\log(p)) \qquad (10.9)$$

for latitudes below 30 degrees, North or South, where p is the desired probability (100 − availability) expressed as a percentage.

ITU rain regions for the Americas, for Europe and Africa, and for Asia are shown in Figures 10.4, 10.5, and 10.6, respectively.

Example 10.1. For the link given in Figure 10.1, determine the depth of a 99.999% fade if the link is located in Florida.

$f = 38.6\,\mathrm{GHz}$; since polarization is not specified, assume horizontal polarization (worst case).

By interpolation, at 38.6 GHz,

$$k = 0.324, \qquad \alpha = 0.95$$

Florida is in ITU rain region N, where 0.01% rain rate is 95 mm/h. Using the four-nines rain rate, the specific attenuation is

$$\mathrm{Atten}_{0.01} = k \cdot RR^{\alpha} = 0.324 \cdot 95^{0.95} = 24.5\,\mathrm{dB}/\mathrm{km}$$

The effective path length is

$$d_0 = 35e^{-0.015 \cdot RR} = 8.417\,\mathrm{km}$$

The distance factor is

$$r = 1/(1 + d/d_0) = 1/(1 + 1.1/8.4) = 0.884$$

So the expected fade depth is

$$d \cdot r \cdot \mathrm{Atten}_{0.01} = 23.8\,\mathrm{dB}$$

Next the predicted attenuation is adjusted to the desired availability using (10.9):

$$\mathrm{Atten} = \mathrm{Atten}_{0.01} \cdot 0.07 \cdot (0.001) - (0.855 + 0.139)\log(0.001)$$
$$\mathrm{Atten} = 34.3\,\mathrm{dB}$$

Since five-nines data are also available, we could use those data to compute attenuation directly. This is not usually done, however, since the ITU model is

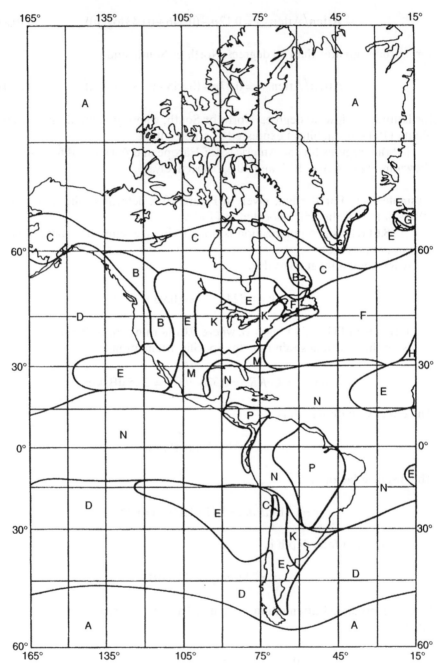

Figure 10.4 ITU rain regions for the Americas (Figure 1 from Ref. 12, courtesy of the ITU.)

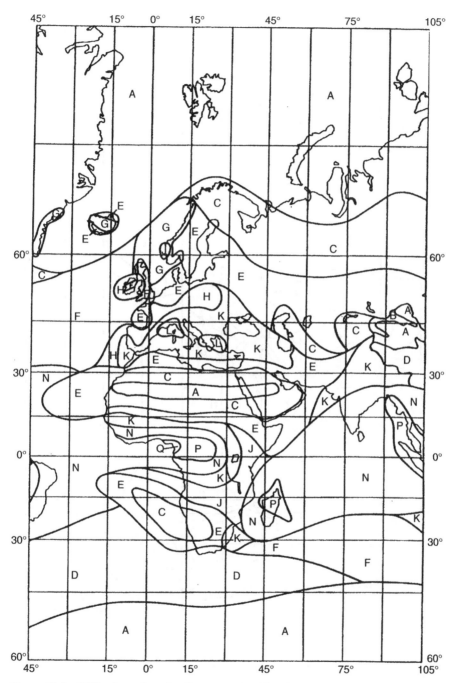

Figure 10.5 ITU rain regions for Europe and Africa. (Figure 2 from Ref. 12, courtesy of the ITU.)

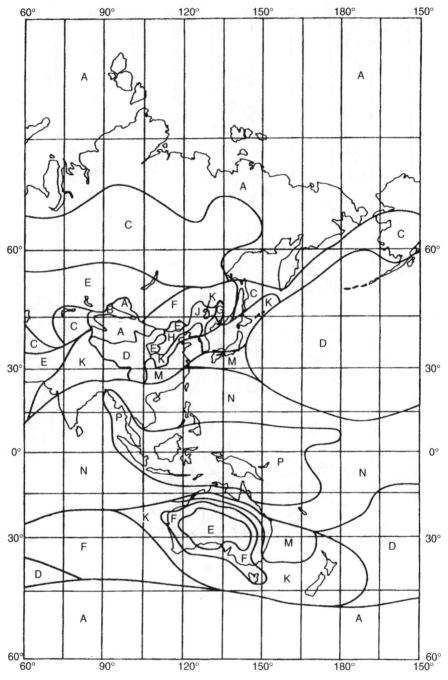

Figure 10.6   ITU rain regions for Asia. (Figure 3 from Ref. 12, courtesy of the ITU.)

based on the four-nines data and the five-nines data has lower confidence since there is less data available. □

10.3.3 The Crane Global Model

The Crane global model is divided into two segments based on distance and the rain rate [15]. The attenuation model is given by

$$Atten = k \cdot RR^{\alpha} (e^{y \cdot \delta} - 1)/y \qquad \text{dB} \quad 0 < d < \delta(RR) \quad \text{km} \qquad (10.10)$$

and

$$Atten = k \cdot RR^{\alpha} \left[\frac{e^{y \cdot \delta(RR)} - 1}{y} + \frac{(e^{z \cdot d} - e^{z \cdot \delta(RR)}) \cdot e^{0.83 - 0.17 \ln(RR)}}{z} \right] \text{dB},$$
$$\delta(RR) < d < 22.5 \text{ km} \qquad (10.11)$$

where $\delta(RR)$ is a function of the rain rate,

$$\delta(RR) = 3.8 - 0.6 \ln(RR) \quad \text{km} \qquad (10.12)$$

d is link distance in km, y is defined as,

$$y = \alpha \cdot \left[\frac{0.83 - 0.17 \ln(RR)}{\delta(RR)} + 0.26 - 0.03 \ln(RR) \right] \qquad (10.13)$$

and

$$z = \alpha \cdot (0.026 - 0.03 \ln(RR)) \qquad (10.14)$$

From the above equations, it can be correctly inferred that the Crane model is only validated for distances up to 22.5 km. Like the ITU model, the Crane model also uses the rain region concept. The Crane rain regions are also labeled alphabetically, but they do not correspond to the ITU regions. Figure 10.7 through Figure 10.9 show the Crane rain regions.

Crane uses different data sets for various probabilities/availabilities, and he does not employ an availability adjustment factor like the ITU model. Table 10.3 gives some of the Crane data for common availabilities, while Table 10A.4 provides the complete set of Crane rain rate values as given in Ref. 16. If the desired availability is not represented in the Crane data, it is possible to (logarithmically) interpolate the given data to estimate the rain rate. While not sanctioned by Dr. Crane, this method provides reasonable information as shown in Figure 10.10 using data for rain region E.

Example 10.2. For the link given in Figure 10.1, determine the depth of a 0.99999 fade if the link is located in Florida.

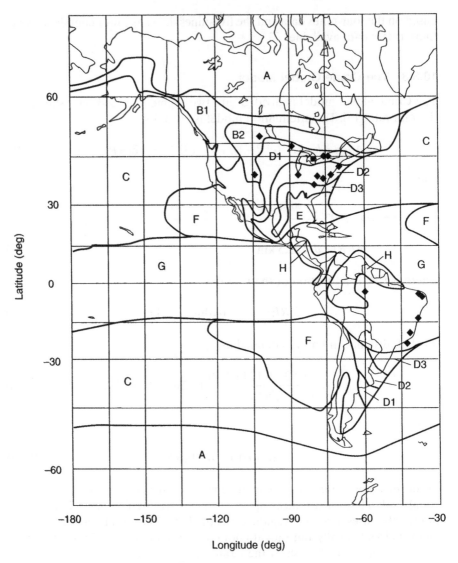

◆ Path locations for rain attenuation measurements

Figure 10.7 Crane rain regions for the Americas. (Figure 3.2 from Ref. 2, courtesy of John Wiley & Sons.)

$f = 38.6\,\text{GHz}$; since polarization is not specified, assume horizontal polarization (worst case).

By interpolation, at 38.6 GHz,

$$k = 0.324, \qquad \alpha = 0.95$$

(these are same coefficients as used for ITU model).

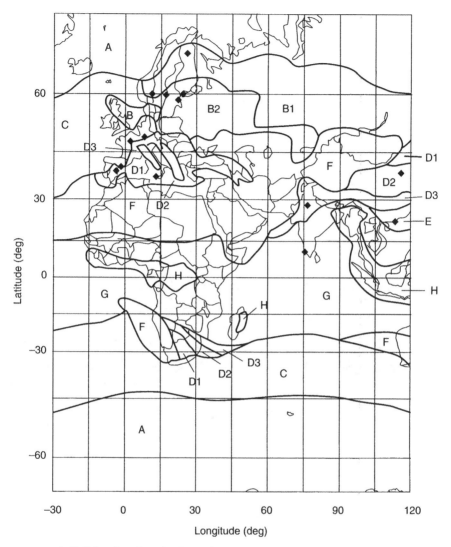

Figure 10.8   Crane rain regions for Europe and Africa. (Figure 3.3 from Ref. 2, courtesy of John Wiley & Sons.)

Florida is in Crane rain region E, where the 0.001% rain rate is 176 mm/h (versus 180 mm/h for ITU region N):

$$\delta(RR) = 3.8 - 0.6\ln(RR) \quad \text{km}$$
$$\delta(176) = 3.8 - 0.6\ln(176) \quad \text{km}$$

Since $d = 1.1$ km, (10.11) applies with

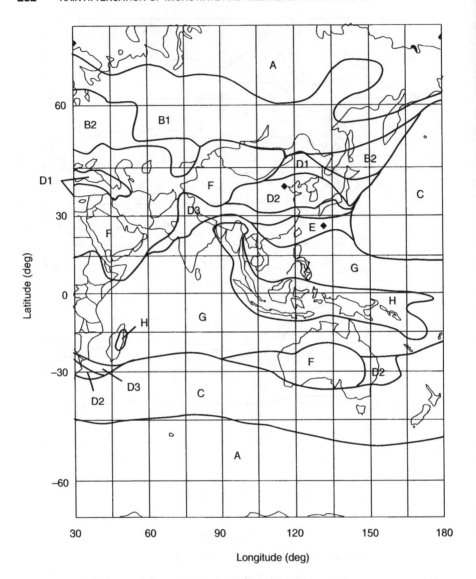

◆ Path locations for rain attenuation measurements

Figure 10.9 Crane rain regions for Asia. (Figure 3.4 from Ref. 2, courtesy of John Wiley & Sons.)

TABLE 10.3   Crane Rain Rate Data

Availability	A	B	B1	B2	C	D1	D2	D3	E	F	G	H
0.99	0.2	1.2	0.8	1.4	1.8	2.2	3.0	4.6	7.0	0.6	8.4	12.4
0.999	2.5	5.7	4.5	6.8	7.7	10.3	15.1	22.4	36.2	5.3	31.3	66.5
0.9999	9.9	21.1	16.1	25.8	29.5	36.2	46.8	61.6	91.5	22.2	90.2	209.3
0.99995	13.8	29.2	22.3	35.7	41.4	49.2	62.1	78.7	112	31.9	118	283.4
0.99999	28.1	52.1	42.6	63.8	71.6	86.6	114.1	133.2	176	70.7	197	542.6

Source: Table 3.1 from Ref. 2, courtesy of John Wiley & Sons.

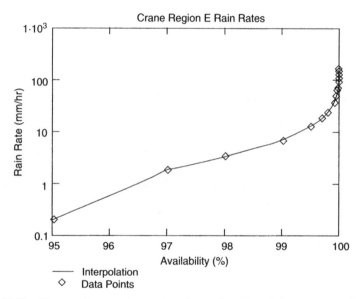

Figure 10.10   Crane rain rate data points for region E and the corresponding logarithmically interpolated curve.

$$y = \alpha \cdot \left[\frac{0.83 - 0.17 \ln(RR)}{\delta(RR)} + 0.26 - 0.03 \ln(RR) \right]$$

and

$$z = \alpha \cdot (0.0226 - 0.03 \ln(RR))$$
$$z = -0.12266$$
$$y = -0.189$$

So,

$$\text{Atten} = 43.9 \text{ dB}$$

Recall that the ITU model predicted 34.3 dB for this link in Example 10.1. ☐

10.3.4 Other Rain Models

There are other rain models available. Some are proprietary and cover limited geographical area. Others are optimized for certain applications. The ITU has adopted an updated model [12], which uses the same attenuation computation, but provides for a much different method for determining the rain statistics. The model uses data files indexed by latitude and longitude to provide a more precise estimate of rain statistics than the rain region concept.

Prior to the Crane global model, Crane developed the two-component model and the revised two-component model, which are both presented in Ref. 2. These models take into account the volume cell and the debris region of a rain event separately and then combine the results. They have, in large part, been superseded by the Crane global model. Both of the two-component models are more involved to implement than the global model.

10.3.5 Rain Attenuation Model Comparison

It is difficult to make generalizations concerning the predicted attenuations from the ITU and Crane global models. Sometimes one or the other will predict more attenuation, sometimes by a significant margin, but on the whole they are fairly consistent. In industry, customers frequently have strong preferences for one model or the other. It is also sometimes the case that a particular model must be used to facilitate comparisons with competing products.

10.3.6 Slant Paths

The details of applying the rain models to slant paths that exit the troposphere are presented in Chapter 11, "Satellite Communications." For terrestrial slant paths, simply using the appropriate value of θ in the expressions for α and k is all that is required. When considering slant paths that exit the troposphere, the statistics of the rain cell height must be incorporated into the model. For heavy rain associated with thunderstorms (convective rain), the rain cells tend to be concentrated, or of limited extent and height. At lower rain rates the rain may exist at much higher elevations and over considerable distances (stratiform rain). The height statistics become important for long paths at high elevation angles and are treated differently than the terrestrial paths.

10.4 THE LINK DISTANCE CHART

The link distance chart is a concise way to convey information on rain fades and margins. The link distance chart consists of a plot of the available fade margin $(G_L - L(d))$ versus link distance [17]. Note that the available fade margin is a monotonically decreasing function with distance due to the free-space loss and any clear-air atmospheric loss. Next, a curve of rain fade depth versus distance is superimposed for each rain region of interest, resulting in a family of curves for the system. The point where the curves intersect defines the maximum link distance for that availability. This is the point where the available margin is exactly equal to the predicted rain-fade depth. Of course the designer may choose to use a different independent variable; for example, if the link distance were fixed, transmitter power, receiver sensitivity, or antenna gain might be used for the independent variable (x axis).

Example 10.3. Consider a 38-GHz terrestrial millimeter-wave link with 180 dB of link gain using vertical polarization. What are the maximum five-nines (99.999%) availability link distances for Arizona and Florida?

Figure 10.11 is the link distance chart for this system using the ITU model, while Figure 10.12 is the link distance chart using the Crane global model. Noting that Florida is in ITU region N and Crane region E and that Arizona is in ITU region E and Crane region F, the results are summarized in Table 10.4. Both link distance charts use the adjusted link gain (i.e., include atmospheric loss), so that the plotted margin is available for rain. It is interesting to note that in one case (Florida) the ITU model is more conservative, while the Crane model is more conservative in Arizona. This relationship between these two models at this geographical location may not hold for all frequencies or availabilities/distances.

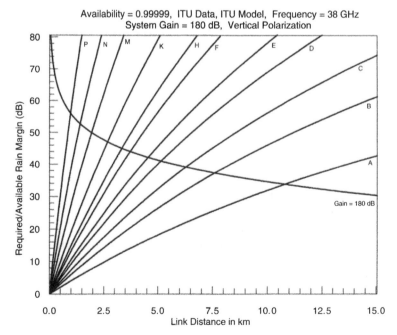

Figure 10.11 Link distance chart using the ITU model and data.

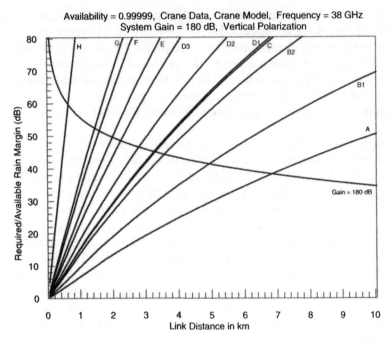

Figure 10.12 Link distance chart using the Crane global model and data.

TABLE 10.4 Link Distance Results for Arizona and Florida

Model	Arizona	Florida
ITU	4.56 km	1.42 km
Crane	3.31 km	1.60 km

It is important to note that the Crane global model is only validated up to 22.5-km link distance, whereas the latest ITU model is valid to 60 km. Since the Crane global model is a curve fit, unpredictable results will be obtained if distances greater than 22.5 km are used.

There are other possible sources of link outages besides rain fades, including hardware failures, interference, physical obstructions, and changes in antenna aim point. These should all be considered when planning availability. It is noteworthy that four- or five-nines availability requires large data sets to achieve high confidence on the statistics. Also note that four-nines is approximately 53 minutes per year of link outage and five-nines is approximately 5 minutes per year.

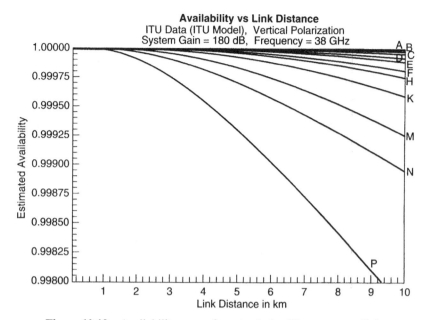

Figure 10.13   Availability curve for a typical millimeter-wave link.

10.5 AVAILABILITY CURVES

The availability adjustment factor in the ITU model may be used to produce curves of availability versus link distance for a given link gain. The available margin for each distance is used in the appropriate adjustment equation, (10.8) or (10.9), depending upon the latitude.

$$\text{Avail}\_\text{margin}/\text{Atten}_{0.01} = 0.12p - (0.546 + 0.043\log(p))$$

This equation can then be iteratively solved for p and the availability is $100 - p$. This is valuable in answering the question of how much availability can be achieved with a given system and link distance. Figure 10.13 shows an example of an availability curve for the same vertically polarized, 180-dB system used earlier. Note that the curves do not exceed five-nines, since that is the confidence limit of the ITU model (and the Crane model as well).

10.6 OTHER PRECIPITATION

Other forms of precipitation such as snow, sleet, and fog may also cause some attenuation of an electromagnetic wave, depending upon the frequency. The varying nature of sleet makes it difficult to definitively model. The easiest approach is to simply treat the sleet as if it were rainfall. This may result in slightly overestimating the expected attenuation. A similar situation exists for snowfall. At frequencies below the visible light range, the primary factor in snow attenuation will be the moisture content. It is likely that the attenuation to either sleet or snow will never reach the same level as the attenuation for rainfall at that same location. Therefore, planning the communications link based on the expected rainfall will likely provide more than enough margin for snow and sleet.

The effect of fog and clouds on electromagnetic waves becomes progressively more significant as the frequency increases. The ITU provides a model for cloud and fog attenuation between 5 and 200 GHz [18]. The model consists of computing a specific attenuation based on the liquid water density in the atmosphere and a specific attenuation coefficient, K_l, that is a function of frequency and temperature. Figure 10.14 shows plots of the specific attenuation coefficient versus frequency for several different temperatures. The reference also provides the equation for reproducing this plot in whatever detail is required. The expression for the specific attenuation due to fog or clouds is

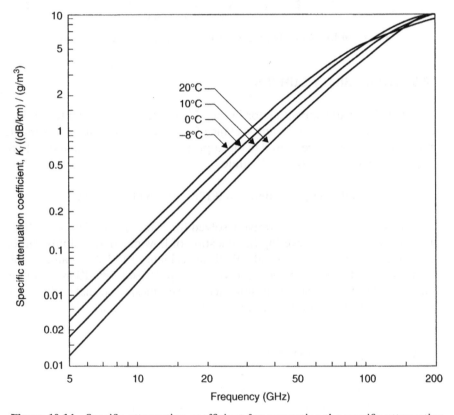

Figure 10.14 Specific attenuation coefficient for computing the specific attenuation due to clouds or fog. (Figure 1 from Ref. 18, courtesy of the ITU.)

$$\gamma_c = K_l M \qquad dB/km \qquad (10.15)$$

where

γ_c is the specific attenuation within the fog or cloud

K_l is the specific attenuation coefficient from Figure 10.13

M is the liquid water vapor density in the fog or cloud

For cloud attenuation, the curve for 0°C should be used. The values for liquid water vapor density are ideally determined by local measurement. Historical data are available from the ITU in data files and are summarized in the maps in Ref. 18. Typical values of water vapor density are 0.05 g/m³ for medium fog (300-m visibility) and 0.5 g/m³ for thick fog (50-m visibility).

Example 10.4. What is the expected attenuation due to fog on a 15-km path for a 30-GHz communication system? Assume that the ambient temperature is 10°C and that the fog is heavy.

The specific attenuation at 30 GHz and 10°C is 0.63 (dB/km)/(g/m³). For heavy fog, use $M = 0.5$ g/m³. The predicted specific attenuation is computed to be 0.315 dB/km. On a 15-km path, the overall attenuation due to the fog will therefore be 4.7 dB. □

10.7 CROSS-POLARIZATION EFFECTS

In addition to the attenuation of electromagnetic waves, rain and other precipitation tend to cause depolarization of the wave. This can be a concern when orthogonal polarizations are used for frequency re-use. It is possible to achieve between 20 and 35 dB of link-to-link isolation by using orthogonal polarizations. The isolation will be reduced by rain and by ground and atmospheric multipath. For this reason, few systems depend heavily on cross-polarization isolation for link isolation and frequency re-use. The ITU [8] gives a procedure for estimating the reduction is cross-polarization isolation in rain and multipath.

10.8 SUMMARY

When high availability is of interest, key system parameters must be known with confidence. While rain is the principal limitation of millimeter-wave link availability, other factors such as hardware failures should not be ignored. Once the rain availability requirement is determined, rain models can be used to determine the fade depth that will not be exceeded with that probability.

Rain fade probabilities can be calculated using the ITU, Crane global, or other models. While the Crane global and ITU models are distinct, they used much of the same raw data in their development and they provide reasonably consistent results. The ITU model is validated for links as long as 60 km, whereas the Crane model is validated up to 22.5 km.

The *adjusted link gain* of a communication link, minus the free-space loss, gives the available rain fade margin. The distance at which the predicted rain fade equals the rain fade margin is the maximum link distance for the chosen availability. Using the ITU model, it is also possible to solve for availability versus distance.

Note that it is possible to use the Crane model with ITU rain rates and vice versa, but there is no significant advantage to doing so other than perhaps to compare the models with identical data or compare the data using identical models.

REFERENCES

1. Special issue on Ka-band propagation effects on earth-satellite links, *Proceedings of the IEEE*, Vol. 85, No. 6, June 1997.

2. R. K. Crane, *Electromagnetic Wave Propagation Through Rain*, John Wiley & Sons, New York, 1996.

3. R. K. Crane, *Propagation Handbook for Wireless Communication System Design*, CRC Press LLC, Boca Raton, FL, 2003.

4. R. K. Crane, Prediction of attenuation by rain, *IEEE Transactions on Communications*, Vol. 28, No. 9, September, 1980, pp. 1717–1733.

5. D. A. de Wolf, H. W. J. Russchenberg, and L. P. Ligthart, Simplified analysis of line-of-sight propagation through rain at 5–90 GHz, *IEEE Transactions on Antennas and Propagation*, Vol. 40, No. 8, Augurst, 1992, pp. 912–919.

6. D. A. de Wolf and A. J. Zwiesler, Rayleigh–Mie approximation for line-of-sight propagation through rain at 5–90 GHz, *IEEE Transactions on Antennas and Propagation*, Vol. 44, No. 3, March, 1996, pp. 273–279.

7. D. A. de Wolf, H. W. J. Russchenberg, and L. P. Ligthart, Attenuation of co- and cross-polarized electric fields of waves through a layer of dielectric spheroids, *IEEE Transactions on Antennas and Propagation*, Vol. 39, No. 2, February, 1991, pp. 204–210.

8. ITU-R Recommendations, *Propagation data and prediction methods required for the design of terrestrial line-of-sight systems*, ITU-R P.530-9, Geneva, 2001.

9. J. Goldhirsh, B. H. Musiani, and W. J. Vogel, Cumulative fade distributions and frequency scaling techniques at 20 GHz from the advanced communications technology satellite and at 12 GHz from the digital satellite system, *Proceedings of the IEEE*, Vol. 85, No. 6, June, 1997, pp. 910–916.

10. ITU-R Recommendations, *Specific attenuation model for rain for use in prediction methods*, ITU-R P.838-1, Geneva, 1999.

11. ITU-R Recommendations, *Characteristics of precipitation for propagation modelling*, ITU-R P.837-1, Geneva, 1994.

12. ITU-R Recommendations, *characteristics of precipitation for propagation modelling*, ITU-R P.837-3, Geneva, 2001.

13. T. Pratt, C. W. Bostian, and J. E. Allnutt, *Satellite Communications*, 2nd ed., John Wiley & Sons, New York, 2003, p. 318.

14. T. Pratt, C. W. Bostian, and J. E. Allnutt, *Satellite Communications*, 2nd ed., John Wiley & Sons, New York, 2003, pp. 510–511.

15. R. K. Crane, *Electromagnetic Wave Propagation Through Rain*, John Wiley & Sons, New York, 1996, p. 147.

16. R. K. Crane, *Electromagnetic Wave Propagation Through Rain*, John Wiley & Sons, New York, 1996, p. 114.

17. J. S. Seybold, Performance prediction for fixed microwave data links, *RF Design Magazine*, May 2002, pp. 58–66.

18. ITU-R Recommendations, *Attenuation due to clouds and fog*, IRU-R P.840-3, Geneva, 1999.

EXERCISES

1. What is the reference rainfall rate and the specific attenuation due to rain for a 99.99% availability link in Chicago, IL? Compute for vertical, horizontal, and circular polarization.

2. Compute the expected 99.9% fade depth for a 28-GHz link operating in Atlanta, Georgia. Assume that the link distance is 2 km. Compute for both vertical and horizontal polarization using the ITU model.

3. Compute the expected 99.995% fade depth for a 20-GHz link operating in Orlando, Florida. Assume that the link distance is 2 km. Compute for both vertical and horizontal polarization using the Crane model.

4. What is the 99.999% rain fade depth for a 28-GHz point-to-point link operating in Seattle, WA? Assume that the path is 4 km and horizontal and that circular polarization is used.
   (a) Using the ITU model
   (b) Using the Crane model

5. What is the 99% rain fade depth for a 42-GHz, circularly polarized link operating in Boston, MA if the link distance is 2 km and the path is sloped at 3 degrees?

APPENDIX 10A: DATA FOR RAIN ATTENUATION MODELS

TABLE 10A.1 Linear Regression Coefficients for Determination of Specific Attenuation Due to Rain as a Function of Polarization

Frequency (GHz)	k_H	k_V	α_H	α_V
1	0.0000387	0.0000352	0.912	0.880
2	0.000154	0.000138	0.963	0.923
4	0.000650	0.000591	1.121	1.075
6	0.00175	0.00155	1.308	1.265
7	0.00301	0.00265	1.332	1.312
8	0.00454	0.00395	1.327	1.310
10	0.0101	0.00887	1.276	1.264
12	0.0188	0.0168	1.217	1.200
15	0.0367	0.0335	1.154	1.128
20	0.0751	0.0691	1.099	1.065
25	0.124	0.113	1.061	1.030
30	0.187	0.167	1.021	1.000
35	0.263	0.233	0.979	0.963
40	0.350	0.310	0.939	0.929
45	0.442	0.393	0.903	0.897
50	0.536	0.479	0.873	0.868
60	0.707	0.642	0.826	0.824
70	0.851	0.784	0.793	0.793
80	0.975	0.906	0.769	0.769
90	1.06	0.999	0.753	0.754
100	1.12	1.06	0.743	0.744
120	1.18	1.13	0.731	0.732
150	1.31	1.27	0.710	0.711
200	1.45	1.42	0.689	0.690
300	1.36	1.35	0.688	0.689
400	1.32	1.31	0.683	0.684

Source: Table 1 from Ref. 10, courtesy of the ITU.

TABLE 10A.2 Interpolated Regression Coefficients for 1–40 GHz

f(GHz)	k_H	α_H	k_V	α_V
1	3.87×10^{-5}	0.912	3.52×10^{-5}	0.88
2	1.54×10^{-4}	0.963	1.38×10^{-4}	0.923
3	3.576×10^{-4}	1.055	3.232×10^{-4}	1.012
4	6.5×10^{-4}	1.121	5.91×10^{-4}	1.075
5	1.121×10^{-3}	1.224	1.005×10^{-3}	1.18
6	1.75×10^{-3}	1.308	1.55×10^{-3}	1.265
7	3.01×10^{-3}	1.332	2.65×10^{-3}	1.312
8	4.54×10^{-3}	1.327	3.95×10^{-3}	1.31
9	6.924×10^{-3}	1.3	6.054×10^{-3}	1.286
10	0.01	1.276	8.87E-3	1.264
11	0.014	1.245	0.012	1.231
12	0.019	1.217	0.017	1.2
13	0.024	1.194	0.022	1.174
14	0.03	1.173	0.027	1.15
15	0.037	1.154	0.034	1.128
16	0.043	1.142	0.039	1.114
17	0.05	1.13	0.046	1.101
18	0.058	1.119	0.053	1.088
19	0.066	1.109	0.061	1.076
20	0.075	1.099	0.069	1.065
21	0.084	1.091	0.077	1.057
22	0.093	1.083	0.085	1.05
23	0.103	1.075	0.094	1.043
24	0.113	1.068	0.103	1.036
25	0.124	1.061	0.113	1.03
26	0.135	1.052	0.123	1.024
27	0.147	1.044	0.133	1.017
28	0.16	1.036	0.144	1.011
29	0.173	1.028	0.155	1.006
30	0.187	1.021	0.167	1
31	0.201	1.012	0.179	0.992
32	0.216	1.003	0.192	0.985
33	0.231	0.995	0.205	0.977
34	0.247	0.987	0.219	0.97
35	0.263	0.979	0.233	0.963
36	0.279	0.971	0.247	0.956
37	0.296	0.962	0.262	0.949
38	0.314	0.954	0.278	0.942
39	0.332	0.947	0.294	0.935
40	0.35	0.939	0.31	0.929

TABLE 10A.3 ITU Rainfall Rates[a] for Different Probabilities and Rain Regions

Percentage of Time (%)	A	B	C	D	E	F	G	H	J	K	L	M	N	P	Q
1.0	<0.1	0.5	0.7	2.1	0.6	1.7	3	2	8	1.5	2	4	5	12	24
0.3	0.8	2	2.8	4.5	2.4	4.5	7	4	13	4.2	7	11	15	34	49
0.1	2	3	5	8	6	8	12	10	20	12	15	22	35	65	72
0.03	5	6	9	13	12	15	20	18	28	23	33	40	65	105	96
0.01	8	12	15	19	22	28	30	32	35	42	60	63	95	145	115
0.003	14	21	26	29	41	54	45	55	45	70	105	95	140	200	142
0.001	22	32	42	42	70	78	65	83	55	100	150	120	180	250	170

[a] Rainfall intensity exceeded (mm/h).

Source: Table 1 from Ref. 11, courtesy of the ITU.

TABLE 10A.4   Crane Rain Rates for Different Probabilities and Rain Regions

Percent of Year	A RR	B RR	B1 RR	B2 RR	C RR	D1 RR	D2 RR	D3 RR	E RR	F RR	G RR	H RR
5	0	0.2	0.1	0.2	0.3	0.2	0.3	0	0.2	0.1	1.8	1.1
3	0	0.3	0.2	0.4	0.6	0.6	0.9	0.8	1.8	0.1	3.4	3.3
2	0.1	0.5	0.4	0.7	1.1	1.2	1.5	2	3.3	0.2	5	5.8
1	0.2	1.2	0.8	1.4	1.8	2.2	3	4.6	7	0.6	8.4	12.4
0.5	0.5	2	1.5	2.4	2.9	3.8	5.3	8.2	12.6	1.4	13.2	22.6
0.3	1.1	2.9	2.2	3.4	4.1	5.3	7.6	11.8	18.4	2.2	17.7	33.1
0.2	1.5	3.8	2.9	4.4	5.2	6.8	9.9	15.2	24.1	3.1	22	43.5
0.1	2.5	5.7	4.5	6.8	7.7	10.3	15.1	22.4	36.2	5.3	31.3	66.5
0.05	4	8.6	6.8	10.3	11.5	15.3	22.2	31.6	50.4	8.5	43.8	97.2
0.03	5.5	11.6	9	13.9	15.6	20.3	28.6	39.9	62.4	11.8	55.8	125.9
0.02	6.9	14.6	11.3	17.6	19.9	25.4	34.7	47	72.2	15	66.8	152.4
0.01	9.9	21.1	16.1	25.8	29.5	36.2	46.8	61.6	91.5	22.2	90.2	209.3
0.005	13.8	29.2	22.3	35.7	41.4	49.2	62.1	78.7	112	31.9	118	283.4
0.003	17.5	36.1	27.8	43.8	50.6	60.4	75.6	93.5	130	41.4	140.8	350.3
0.002	20.9	41.7	32.7	50.9	58.9	69	88.3	106.6	145.4	50.4	159.6	413.9
0.001	28.1	52.1	42.6	63.8	71.6	86.6	114.1	133.2	176	70.7	197	542.6

Source: Table 3.1 from Ref. 2, courtesy of John Wiley & Sons.

CHAPTER 11

Satellite Communications

11.1 INTRODUCTION

There are numerous applications of satellite systems, including remote sensing (SAR, IR, visual), communication links for telephone, aircraft, and military purposes, mobile communications such as satphones, Iridium, and so on, direct broadcast satellite TV, navigation systems such as GPS, and communication eavesdropping for intelligence collection. While each of these objectives have their own unique requirements, all satellite systems share certain characteristics and operate in the same kind of environment. It is also essential that all man-made satellites be able to communicate with their terrestrial control stations regardless of their primary mission.

Satellite communication links entail a signal traversing the entire atmosphere and therefore can suffer significantly due to atmospheric factors, depending upon the frequency. In addition, the distances involved are considerable, so the free-space loss is much larger than it is for terrestrial links. The costs of reaching orbit dictates that satellite aperture size and transmit power are at a premium, making reliable link design a challenge.

In this chapter, many of the channel impairments that have been previously discussed are revisited and it is shown how they are applied to satellite communication links. First the various types of satellite orbits and their characteristics are briefly examined, followed by determination of the range to the satellite. Computation of the actual line-of-sight or slant-range distance to the satellite is somewhat complicated by the curvature of the earth. Once the distance to the satellite has been determined and the free-space loss calculated, other link impairments must be considered. These include atmospheric attenuation (gaseous absorption, rain, clouds, and fog), and ionospheric effects. A considerable amount of time is spent on the atmospheric absorption and rain attenuation, even though they have been covered in earlier chapters. This is for two reasons: (a) At the frequencies of interest for most satellite links, rain and atmospheric attenuation can be significant, and (b) The computation of these losses is somewhat different than for the terrestrial case.

Introduction to RF Propagation, by John S. Seybold
Copyright © 2005 by John Wiley & Sons, Inc.

The unique aspects of antennas for satellite communications are discussed, including sky noise and radome effects. It is also imperative that a satellite link analysis takes into account the fact that many of the atmospheric losses are absorptive in nature and therefore raise the apparent receiver noise floor, by effectively increasing the sky temperature. This often-overlooked effect can cause errors of several dB in computing link performance for low-noise receivers.

Scintillation is a form of fast fading that can be attributed to multipath from variations in the ionosphere or the troposphere. While the mechanisms of these two types of scintillation are different, the effect on the communication link is similar: A variation in signal strength that may cause tracking, detection, and AGC problems. Tropospheric scintillation can be problematic for nearly horizontal paths, but is rarely a concern for satellite paths. Tropospheric scintillation is due to variations in temperature and humidity and can apply to frequencies as high as 30 GHz. Since the lower atmosphere is usually horizontally stratified, this effect is most pronounced at low elevation angles [1]. Ionospheric scintillation only affects signals below 10 GHz, but is not sensitive to elevation angle. Other ionospheric phenomena that are of concern for satellite links include propagation delay, dispersion, and Faraday rotation.

Foliage can also be a significant signal attenuator. In some cases the losses due to foliage are so large that the designer simply treats them as blockages. In the case of a mobile system, this results in brief outages or reduced data throughput. For static systems, this approach renders some locations unsuitable. A prime example of this is that North American homes with large trees blocking their line of sight to the southwestern sky cannot receive Direct TV.

11.2 SATELLITE ORBITS

Satellite orbits are loosely grouped into four categories: low earth orbit (LEO), medium earth orbit (MEO), geosynchronous orbit (GEO), and high earth orbit (HEO). Figure 11.1 shows the relationships among the different orbits [2].

The corresponding orbital altitudes (above the earth's surface) are

LEO, 500–900 km
MEO, 5000–12,000 km
GEO, 36,000 km*
HEO, 50,000 km

These definitions are not universal and may be applied rather casually. HEO satellites have an elliptical orbit, which may be very near the earth at its low

---

* More precisely, 35,786 km from Ref. 2, pp. 19 and 35.

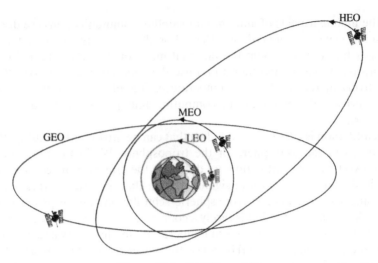

Figure 11.1   Satellite orbits.

point (perigee) and will have a period greater than one sidereal day (23.9344 h, one earth rotation). HEO is often used for scientific applications.

Geosynchronous satellite orbits have an orbital period of exactly one sidereal day and move about in a relatively small angular window, depending upon the orbital inclination. If a geosynchronous orbit has an inclination angle of zero relative to the equator, then the satellite will remain directly above the earth's equator and appear virtually stationary from the earth's surface. This special case of geosynchronous orbit is called *geostationary orbit* (GSO). GEO (and GSO) satellites have little or no relative motion, while all others pass through the sky and produce significant Doppler shift on any electromagnetic waves that are transmitted to or received from the earth.

With the exception of geostationary satellites, the free-space loss will vary with time as the satellite orbits the earth. The direction and magnitude of the Doppler shift will also vary with time as the satellite passes overhead. The Doppler shift depends upon the frequency of operation and is given by

$$f_d = \frac{v_r f_0}{c} \tag{11.1}$$

where f_0 is the center frequency and v_r is the relative velocity between the satellite and the earth terminal.

While geostationary satellites have the advantage of minimal Doppler shift, the corresponding higher orbital altitude produces a much longer time delay and greater free-space loss than the lower orbits. Geosynchronous satellites, while not stationary, tend to move at a lower rate (relative to the earth) than LEOs, MEOs, and HEOs, which relaxes the Doppler and angular tracking requirements. The time delay can be a concern for satellite communications, especially for HEOs and GEOs. For GEO orbits where the round trip is 72,000 km, the corresponding delay is approximately 240 ms. The greater path loss and signal delay for GEO satellites relative to LEOs and MEOs are somewhat offset by the reduced tracking requirements.

11.3 SATELLITE OPERATING FREQUENCY

Satellite communications occur from frequencies of tens of MHz, to 40 GHz and beyond. The higher frequencies permit much greater bandwidth, which permits greater data flow and can reduce power requirements. This is a significant consideration as bandwidth demands increase and the electromagnetic spectrum becomes more crowded. Higher frequencies also permit greater antenna gain for a given aperture size. Conversely, lower frequencies have less propagation losses (gaseous, clouds, rain) as well as less free-space loss and are better able to penetrate buildings and foliage. Any regulatory restrictions in the geographic areas covered by the satellite must also be taken into account. Thus the choice of operating frequencies includes many considerations.

A recent trend has been toward ka-band satellites, which use 30-GHz uplink and 20.2-GHz downlink frequencies. The wide separation in frequency makes transmit-to-receive isolation easier to attain. The reason for this migration to higher frequencies is spectral crowding, antenna aperture size restrictions, and an improved understanding of rain attenuation in the ka band due in large part to the ACTS experiment [3]. Higher frequency (and satellite power) is what has made direct broadcast television possible. It was not that long ago that having satellite TV required a fairly large C-band antenna in the backyard.

Literally on the opposite end of the spectrum are the amateur radio satellites (more correctly, satellite payloads because they usually piggy back on a larger satellite). The amateur satellites often use 140-MHz uplinks and 30-MHz downlinks, or 440-MHz uplinks and 145-MHz downlinks. The primary motivation for these frequency selections is the wide availability and low cost of amateur equipment in these bands. This permits the satellite to service the largest possible segment of the amateur community. All amateur satellites are LEOs because the path loss of the higher orbits would make low-cost communications very difficult. By using LEOs, low-power and omnidirection antennas can be used to communicate (albeit briefly) from one ground station to another. Leo satellites also permit worldwide usage.

11.4 SATELLITE PATH FREE-SPACE LOSS

The largest path loss element in a satellite link is of course the free-space loss, which is a function of the distance between the ground station and the satellite and the frequency or wavelength. The distance to the satellite is a

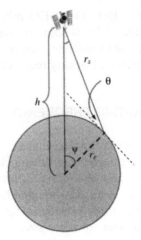

Figure 11.2   Satellite slant-path geometry.

function of the satellite altitude and elevation angle at which it is viewed (due to the earth's curvature) [4]. Figure 11.2 shows the geometry, where

r_e is the earth's radius (6378 km)
h is the satellite height *above the center* of the earth
θ is the elevation angle at which the satellite appears

The law of sines provides the relationships needed to solve for r_s:

$$\frac{h}{\sin\left(\dfrac{\pi}{2}+\theta\right)} = \frac{r_s}{\sin(\psi)} = \frac{r_e}{\sin\left(\dfrac{\pi}{2}-\theta-\psi\right)}$$

so

$$r_s = \frac{h\sin(\psi)}{\sin\left(\dfrac{\pi}{2}+\theta\right)} \tag{11.2}$$

where the central angle is given by

$$\psi = \cos^{-1}\left(\frac{r_e}{h}\sin\left(\frac{\pi}{2}+\theta\right)\right)-\theta \tag{11.3}$$

If the elevation angle, θ, is not known, the central angle can be found using the expression [4]

$$\cos(\psi) = \cos(L_e)\cos(L_s)\cos(l_s - l_e) + \sin(L_e)\sin(L_s) \qquad (11.4)$$

where

L_e is the earth station latitude
l_e is the earth station longitude
L_s is the subsatellite latitude
l_s is the subsatellite longitude

With the central angle known, the slant-path distance and elevation angle are readily found using

$$r_s = h\sqrt{1 + \left(\frac{r_e}{h}\right)^2 - 2\left(\frac{r_e}{h}\right)\cos(\psi)} \qquad (11.5)$$

and

$$\cos(\theta) = \frac{h\sin(\psi)}{r_s} \qquad (11.6)$$

Alternatively, one may use the slant range to the satellite (if known) and its height, to solve for the elevation angle using the law of cosines. Note that the actual earth radius is used in this calculation, not the 4/3's approximation [4]. The 4/3's earth approximation is only appropriate for terrestrial paths that are nearly horizontal and simply aids in finding the radio horizon. The effect of refraction on the elevation angle is negligible for slant paths with elevation angles greater than 2–3 degrees [5].

In the case of a GEO satellite, the height is fixed, so the communication distance becomes a function of the elevation angle, which is, in turn, a function of the location (latitude, longitude) of the earth terminal. With $h_{GEO} = 42,242\,\text{km}$,

$$r_s = \frac{h_{GEO}\sin\left(\cos^{-1}\left(\frac{r_e}{h_{GEO}}\sin\left(\frac{\pi}{2}+\theta\right)\right)-\theta\right)}{\sin\left(\frac{\pi}{2}+\theta\right)} \qquad (11.7)$$

This distance is then used in the Friss free-space loss equation to determine the free-space path loss to the satellite.

Example 11.1. What is the elevation angle and path distance between a geostationary satellite and its earth terminal if the earth terminal is located at 20 degrees latitude? You may assume that the satellite and earth station are at the same longitude, –90 degrees.

The earth station is at 20 degrees latitude and the geostationary satellite is by definition at 0 degrees latitude, so $\psi = 20$ degrees in Figure 11.2. It is known that $h = 42{,}242\,\text{km}$ and $r_e = 6378\,\text{km}$, so

$$r_s = h\sqrt{1 + \left(\frac{r_e}{h}\right)^2 - 2\left(\frac{r_e}{h}\right)\cos(\psi)}$$

implies that $r_s = 36{,}314\,\text{km}$. Using this value for r_s, the elevation angle can now be determined from

$$\cos(\theta) = \frac{h\sin(\psi)}{r_s}$$

which yields $\theta = 66.6$ degrees. $\Box$

11.5 ATMOSPHERIC ATTENUATION

When considering atmospheric absorption on a slant path, the variation of the atmosphere with altitude must be considered. While temperature, pressure, and water vapor content may be constant for terrestrial paths over limited distances, a slant path will cross a continuum of conditions as it passes up through the atmosphere. Mathematically, this can be treated as if the atmosphere were made up of discrete layers. When the slant path completely exits the troposphere such as for an earth-to-space link, the attenuation estimate may be based on the entire columnar oxygen and water vapor content of the atmosphere. In this case, the total length of the path is no longer a factor in computing atmospheric loss, since only the length of the segment that is within the atmosphere contributes to the attenuation.

The ITU provides a detailed model, called the *line-by-line* model for determining the specific attenuation of the atmosphere [6]. The most accurate estimate of slant-path attenuation is obtained by integrating the results of the ITU line-by-line specific attenuation over altitude [6]. Because of its complexity, it is not often employed in practice, but instead a curve fit, also provided in Ref. 6, is used to predict the specific attenuation with less than a 15% error. For most applications, this simplified calculation presented in Ref. 6 is sufficient. Figure 11.3 shows a plot of the slant-path attenuation versus frequency for several elevation angles that was generated using the simplified approach. The simplified approach uses standard atmosphere specific attenuations, equivalent heights, and the elevation angle of the path to provide an estimate of total attenuation that is within 10% of the more detailed computation. This method is valid for elevation angles between 5 and 90 degrees, ground station altitudes between 0 and 2 km, and frequencies that are more than 500 MHz away from the centers of the resonance lines (shown in Figure 11.4). The path attenuation is given by

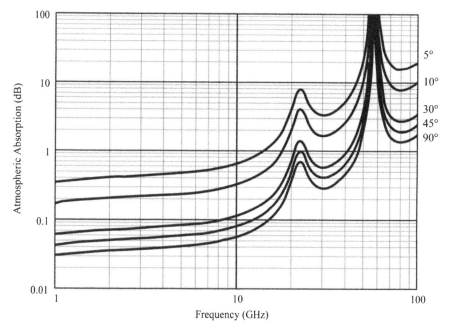

Figure 11.3 Atmospheric absorption for an earth–space path.

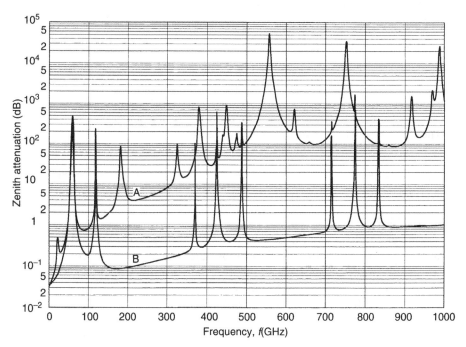

Figure 11.4 Zenith attenuation due to atmospheric gases. Curve A: Mean global reference atmosphere (7.5 g/m³ at sea level). Curve B: Dry atmosphere. (Figure 3 from Ref. 6, courtesy of the ITU.

$$A = \frac{h_0 \gamma_0 + h_w \gamma_w}{\sin(\theta)} \quad \text{dB} \qquad (11.8)$$

The specific attenuations for oxygen and water vapor (γ_0 and γ_w, respectively) are the same values used for terrestrial links from Figure 10.1, θ is the elevation angle, and h_0 and h_w are the equivalent oxygen and water vapor heights, respectively.

For $1 \leq f \leq 56.7\,\text{GHz}$,

$$h_0 = 5.386 - 3.32734 \times 10^{-2} f + 1.87185 \times 10^{-3} f^2 - 3.52087 \times 10^{-5} f^3$$
$$+ \frac{83.26}{(f - 60)^2 + 1.2} \quad \text{km}$$

For $56.7 < f < 63.3\,\text{GHz}$,

$$h_0 = 10\,\text{km}$$

For $63.3 \leq f < 98.5\,\text{GHz}$,

$$h_0 = f \left[\frac{0.039581 - 1.19751 \times 10^{-3} f + 9.14810 \times 10^{-6} f^2}{1 - 0.028687 f + 2.07858 \cdot 10^{-4} f^2} \right] + \frac{90.6}{(f - 60)^2} \quad \text{km}$$

For $98.5 \leq f \leq 350\,\text{GHz}$,

$$h_0 = 5.542 - 1.76414 \times 10^{-3} f + 3.05354 \times 10^{-6} f^2 + \frac{6.815}{(f - 118.75)^2 + 0.321} \quad \text{km}$$

The equivalent water vapor height is given by

$$h_w = 1.65 \left\{ 1 + \frac{1.61}{(f - 22.23)^2 + 2.91} + \frac{3.33}{(f - 183.3)^2 + 4.58} + \frac{1.90}{(f - 325.1)^2 + 3.34} \right\} \quad \text{km}$$

for $f \leq 350\,\text{GHz}$.

This method is valid for ground terminal altitudes of 2 km and below. For slant paths to high-altitude aircraft or satellites, the entire troposphere is traversed and the expected attenuation varies only with the geographical location (environmental conditions) and frequency. Therefore the atmospheric loss does not depend on the actual link distance.

Example 11.2. Determine the gaseous absorption on a 20-degree slant path to a satellite if the frequency is 20 GHz. Assume standard atmospheric parameters and that the ground station is at sea level.

The easiest method to determine the expected attenuation is to interpolate between the 10- and 30-degree curves in Figure 11.3. By interpolating, the attenuation can be estimated to be about 1.4 dB for the entire path. □

11.6 IONOSPHERIC EFFECTS

The ionosphere has many effects on electromagnetic waves, which apply primarily below 10 GHz and are pronounced below 1 GHz. The effects of the ionosphere correlate with sun spot activity and the diurnal cycle. Possible effects include: fading, which can be significant (20 dB or more); depolarization or Faraday rotation, time delay; and dispersion (differential time delay). Table 11.1 provides a summary of ionospheric effects for several frequencies at a 30-degree elevation angle.

Faraday rotation is an angular rotation of the polarization vector. For circular polarization, the only impact is to change the angle between the two sets of polarization axes, which can degrade the cross-polarization isolation and increase the cross-polarization loss if the antennas are not perfectly circularly polarized. This is only significant in systems that have a significant axial ratio loss to begin with (see Chapter 3 for more details). The relative insensitivity of circular polarization to Faraday rotation is one reason circular polarization is often used in satellite communications, particularly below 10 GHz. If linear polarization is used, for maximum energy transfer the antenna polarization axes must be aligned, which can be difficult for space communication even without the effect of Faraday rotation.

Propagation delay may be important in some time critical systems such as navigation or GPS. If two carriers are available, however, it is possible to use the $1/f^2$ relationship to estimate the additional propagation delay and remove it. This is done with GPS receivers that use both the L1 and L2 frequencies. This was one of the techniques that military GPS used to outperform civilian GPS. Recent changes to the GPS satellite constellation and operating doctrine have made dual frequency operation and the accompanying delay correction available to civilian systems as well.

It is interesting that most of the refraction values in the table are on the order of minutes and seconds, not degrees. Only at the HF frequencies is the refraction significant enough to bend the rays back to earth. The remaining effects are negligible below 1 GHz, with the exception of scintillation.

11.7 RAIN FADES

As discussed in Chapter 10, hydrometeors can cause attenuation, scattering, and depolarization of electromagnetic waves. Rain cell models such as Crane or ITU are employed to predict the rain fade depth that will occur with a given

TABLE 11.1 Ionospheric Effects for Several Different Frequencies at 30-Degree Elevation Angle

Effect	Frequency Dependence	0.1 GHz	0.25 GHz	0.5 GHz	1 GHz	3 GHz	10 GHz
Faraday rotation	$1/f^2$	30 rotations	4.8 rotations	1.2 rotations	108°	12°	1.1°
Propagation delay	$1/f^2$	25 μs	4 μs	1 μs	0.25 μs	0.028 μs	0.0025 μs
Refraction	$1/f^2$	<1°	<0.16°	<2.4′	<0.6′	<4.2″	<0.36″
Variation in the direction of arrival (rms)	$1/f^2$	20′	3.2′	48″	12″	1.32″	0.12″
Absorption (auroral and/or polar cap)	$\approx 1/f^2$	5 dB	0.8 dB	0.2 dB	0.05 dB	6×10^{-3} dB	5×10^{-4} dB
Absorption (mid-latitude)	$1/f^2$	<1 dB	<0.16 dB	<0.04 dB	<0.01 dB	<0.001 dB	$<1 \times 10^{-4}$ dB
Dispersion	$1/f^3$	0.4 ps/Hz	0.026 ps/Hz	0.0032 ps/Hz	0.0004 ps/Hz	1.5×10^{-5} ps/Hz	4×10^{-7} ps/Hz
Scintillation[a]	See Rec. ITU-R P.531	See Rec. ITU-R P.531	See Rec. ITU-R P.531	See Rec. ITU-R P.531 Peak-to-peak	>20 dB Peak-to-peak	≈10 dB Peak-to-peak	≈4 dB Peak-to-peak

[a] Values observed near the geomagnetic equator during the early night-time hours (local time) at equinox under conditions of high sunspot number.

Source: Table 1 from Ref. 1 courtesy of the ITU.

probability. The depth of a rain fade is correlated with rain rate and highly correlated with the transmit frequency. The rain fade depth, duration, and frequency of occurrence are highly dependent upon geographical location and even at a given location can vary considerably year-to-year. The modeling of rain fades for satellite links is similar to that for terrestrial links, albeit a little more complex, since the model must account for the variation in rain density with altitude. Rain statistics (for terrestrial or satellite links) tend to be cyclostationary on a month-to-month basis, meaning that the statistics for a particular month are consistent from year-to-year, but the statistics from one month to the next are not.

11.7.1 ITU Rain Attenuation Model for Satellite Paths

The amount of fading due to rain is a function of the frequency and is highly correlated with rain rate. By using rain statistics for a given region, it is possible to determine the probability that a given fade depth will be exceeded. The *rain availability* of a communication link is the complement of the probability of the link fade margin being exceeded. The ITU [7] provides global rain statistics by dividing the earth into rain regions and assigning a rain rate to each region along with the probability of that rain rate being exceeded. The ITU model used here employs only the 0.01% rain statistics and then applies an adjustment factor to the predicted rain fade depth for other probabilities.

For ground-to-space and ground-to-airborne links, the statistical height of the rain cell must be taken into account [1, 8]. The assumption is that for ground-to-airborne links, the airborne platform altitude is above the troposphere. When this assumption holds, the slant-path rain attenuation is no longer a function of the link distance. The latest ITU recommendation [8] references a text file of rain cell height data as a function of latitude and longitude. A less accurate, but nonetheless widely employed, method of computing the expected rain cell height as a function of latitude and the rainfall rate is to use the expression from an earlier recommendation [9]. For latitudes less than 36 degrees, the rain cell height is assumed to be 4 km. For higher latitudes, the rain cell height is taken to be

$$h_R = 4 - 0.075(\text{latitude} - 36°) \tag{11.9}$$

In central Florida then, the nominal rain cell height is 3.6 km.

The ITU [1] gives a 10-step procedure for computing the rain attenuation on a satellite path. That procedure is paraphrased here.

Step 1: Determine the rain height, h_R.
Step 2: Determine the length of the slant path that is below the top of the rain cell.
   For $\theta < 5°$,

$$L_{sl} = \frac{2(h_R - h_s)}{\left(\sin^2(\theta) + \frac{2(h_R - h_s)}{R_e}\right)^{1/2} + \sin(\theta)} \quad \text{km} \tag{11.10}$$

Otherwise,

$$L_{sl} = \frac{(h_R - h_s)}{\sin(\theta)} \quad \text{km} \tag{11.11}$$

where h_s is the surface height where the antenna is located.

Step 3: Calculate the horizontal projection of L_{sl},

$$L_G = L_{sl} \cos(\theta) \tag{11.12}$$

Step 4: Find the 0.01% rainfall rate for the location. For Florida, rain region N,

$$R_{0.01} = 95 \text{ mm/h} \tag{11.13}$$

Step 5: Calculate the specific attenuation for the desired frequency, polarization, and rain rate using the interpolated linear regression coefficients that are found from (10.3) and (10.4).

$$\gamma_R = k(R_{0.01})^\alpha \quad \text{dB/km} \tag{11.14}$$

Step 6: Compute the horizontal reduction factor, $r_{0.01}$ for 0.01% of the time.

$$r_{0.01} = \frac{1}{1 + 0.78\sqrt{\frac{L_G \gamma_R}{f}} - 0.38(1 - e^{-2L_G})} \tag{11.15}$$

where L_G is in km and f is in GHz.

Step 7: Calculate the vertical adjustment factor $v_{0.01}$ for 0.01% of the time.

$$\zeta = \tan^{-1}\left(\frac{h_R - h_s}{L_G r_{0.01}}\right) \quad \text{degrees} \tag{11.16}$$

For $\zeta > \theta$,

$$L_R = \frac{L_G r_{0.01}}{\cos(\theta)} \quad \text{km} \tag{11.17}$$

Otherwise,

$$\frac{(h_R - h_s)}{\sin(\theta)} \quad \text{km} \tag{11.18}$$

If latitude < 36 degrees,

$$\chi = 36 \text{ degrees} - |\text{latitude}|$$

Otherwise,

$$\chi = 0$$

Then,

$$v_{0.01} = \cfrac{1}{1 + \sqrt{\sin(\theta)} \left(31 \left(1 - e^{-(\theta/(1+\chi))} \right) \cfrac{\sqrt{L_R \gamma_R}}{f^2} - 0.45 \right)} \qquad (11.19)$$

Step 8: The effective path length is then computed as

$$L_E = L_R v_{0.01} \qquad \text{km} \qquad (11.20)$$

Step 9: The 0.01% rain fade depth is next computed from

$$A_{0.01} = \gamma_R L_E \qquad \text{dB} \qquad (11.21)$$

Step 10: To determine the maximum fade depths for other probabilities in the range from 0.001% to 5%, the availability adjustment factor is applied. If $p \geq 1\%$ or if $|\text{latitude}| \geq 36$ degrees,

$$\beta = 0 \qquad (11.22)$$

If $p < 1\%$ and $|\text{latitude}| < 36$ degrees and $\theta > 25$ degrees,

$$\beta = -0.005(|\text{latitude}| - 36) \qquad (11.23)$$

Otherwise,

$$\beta = -0.005(|\text{latitude}| - 36°) + 1.8 - 4.25 \sin(\theta) \qquad (11.24)$$

The maximum fade is then given by

$$A_p = A_{0.01} \left(\frac{p}{0.01} \right)^{-(0.655 + 0.033 \ln(p) - 0.045 \ln(A_{0.01} - \beta(1-p)\sin(\theta)))} \qquad (11.25)$$

The overall expected attenuation on an earth–space path ends up being a function of the rain rate (availability), frequency, and elevation angle only. Figure 11.5 shows plots of the expected rain attenuation on an earth–space path versus availability, for several elevation angles and operating frequencies using

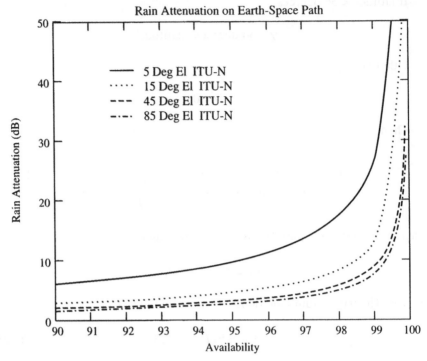

Figure 11.5 Required rain attenuation margin values versus availability for different elevation angles at 30 GHz using the ITU model in rain region N, at 30 degrees latitude.

the ITU model in rain region N. As the required availability increases, so does the amount of rain-fade margin that is needed.

Example 11.3. Consider a geostationary satellite that uses a 30-GHz uplink and a 20-GHz downlink. If the uplink is located in New York City, the downlink is in Miami and the satellite is at 90 degrees longitude, determine the 99.9% fade margin for both the downlink and the uplink if circular polarization is used. What is the overall probability of a communications outage?

First, the satellite and ground station locations are used to determine the elevation angles to the satellite. New York is in rain region K, its latitude is approximately 40 degrees, and its longitude is about 74 degrees. For the uplink, then the earth lat/long parameters are;

$$L_e = 40 \text{ degrees}, \quad l_e = 74 \text{ degrees}$$

the satellite lat/long parameters are;

$$L_s = 0 \text{ degrees}, \quad l_s = 90 \text{ degrees}$$

and, using (11.4),

$$\psi_{UL} = 42.6 \text{ degrees}$$

Next, the uplink slant range is found using (11.5):

$$r_{sUL} = 42,378\sqrt{1+\left(\frac{6,378}{42,378}\right)^2 - 2\left(\frac{6,378}{42,378}\right)^2 \cos(42.6)} = 37,930 \text{ km}$$

The elevation angle is determined using (11.6),

$$\cos(\theta) = \frac{42,378\sin(42.6)}{37,930} = 0.76$$

so

$$\theta_{UL} = 40.9 \text{ degrees}$$

Using information from Table 10A.2, the specific attenuation regression coefficients for the uplink are

$$k_H = 0.187, \qquad \alpha_H = 1.021$$
$$k_V = 0.167, \qquad \alpha_V = 1.0$$

Using the uplink elevation angle of 40.9 degrees and a tilt angle of 45 degrees for circular polarization, the regression coefficients for the uplink are found from (10.3) and (10.4) to be

$$k_{UL} = \frac{k_H + k_V}{2} = 0.177$$

$$\alpha_{UL} = \frac{k_H \alpha_H + k_V \alpha_V}{2k} = 1.011$$

Next, the 10-step procedure is applied.

Step 1: Find the rain height.

$$h_R = 4 - 0.075(\text{latitude} - 36°) = 3.7 \text{ km}$$

Step 2: Determine the length of the slant path that is below the rain height.

$$L_{sl} = \frac{(3.7 - 0)}{\sin(40.9)} = 5.65 \text{ km}$$

Step 3: Calculate the horizontal projection of L_{sl}.

$$L_G = 5.65\cos(40.9) = 4.27 \text{ km}$$

Step 4: Find the 0.01% rainfall rate for rain region K from Table 10.2.

$$R_{0.01} = 42 \text{ mm/h}$$

Step 5: Calculate the specific attenuation for the frequency, polarization, and rain rate.

$$\gamma_R = 0.072(42)^{1.082} = 7.75 \quad \text{dB/km}$$

Step 6: Compute the horizontal reduction factor.

$$r_{0.01} = \cfrac{1}{1+0.78\sqrt{\cfrac{4.27\cdot 4.11}{20}}-0.38(1-e^{-2\cdot 4.27})} = 0.695$$

Step 7: Compute the vertical adjustment factor.

$$\zeta = \tan^{-1}\left(\frac{3.7-0}{4.27\cdot 0.74}\right) = 51.28 \text{ degrees}$$

Since $\zeta > \theta$,

$$L_R = \frac{4.27\cdot 0.74}{\cos(40.9)} = 3.925 \text{ km}$$

Since latitude > 36,

$$\chi = 0$$

and

$$v_{0.01} = \cfrac{1}{1+\sqrt{\sin(40.9°)}\left(31\left(1-e^{-\left(\frac{\pi}{180}\cdot 40.9/(1+0)\right)}\right)\cfrac{\sqrt{3.925\cdot 7.75}}{30^2}-0.45\right)} = 1.267$$

Step 8: Compute the effective path length

$$L_E = 3.925\cdot 1.267 = 4.971 \text{ km}$$

Step 9: Compute the 0.01% fade depth.

$$A_{0.01} = 7.75\cdot 4.971 = 38.52 \text{ dB}$$

Step 10: Compute the fade depth for 0.1% probability.
   For these locations and angles,

$$\beta = 0$$

and the expression for the 0.1% fade is

$$A_{0.1} = 38.52 \left(\frac{0.1}{0.01} \right)^{-(0.655+0.033\ln(0.1)-0.045\ln(38.52)-0(1-0.1)\sin(40.9°))} = 14.8 \, \text{dB}$$

   Thus the uplink rain fade can be expected to exceed 14.8 dB 0.1% of the year.
   Using a similar approach for the downlink, yields

$$L_e = 26 \text{ degrees}, \quad l_e = 80 \text{ degrees}$$
$$L_s = 0 \text{ degrees}, \quad l_s = 90 \text{ degrees}$$

The central angle is

$$\psi_{DL} = 27.7 \text{ degrees}$$

from which the slant range is found,

$$r_{sDL} = 36,850 \, \text{km}$$

and then the elevation angle to the satellite can be determined

$$\theta_{DL} = 57.7 \text{ degrees}$$

The 20-GHz regression coefficients are

$$k_H = 0.075, \quad \alpha_H = 1.099$$
$$k_V = 0.069, \quad \alpha_V = 1.065$$

for circular polarization, the regression coefficients are

$$k = 0.072, \quad \alpha = 1.083$$

The 0.01% attenuation is

$$A_{0.01} = 43.25 \, \text{dB}$$

And finally the 0.1% fade depth is 16.85 dB.
   Thus the downlink rain fade can be expected to exceed 16.85 dB 0.1% of the year. In this case, the uplink and downlink fades are comparable. The

30-GHz uplink has slightly less attenuation even though it has the lower elevation angle. This is due to the more moderate rain region in New York. The rain in Miami is heavy enough to cause the lower-frequency (20-GHz) downlink at the greater elevation angle to still have a slightly larger rain fade.

To determine the overall link availability, the weather in Miami and New York are treated as independent. Furthermore, it is clear that if either end fades out, the link will drop. The probability of the link being out is one minus the probability of both links being operational, which, since they are independent, is simply the product of their probabilities.

$$P_{outage} = 1 - P_{UL}P_{DL} = 1.999 \times 10^{-3} \sim 0.002 \quad \square$$

11.7.2   Crane Rain Attenuation Model for Satellite Paths

The Crane global model [10, 11] is similar to the ITU model in that it also employs the concept of rain regions for determining expected rain rates and that the attenuation is a function of the range rate. It differs in that the data are slightly different and the attenuation model is considerably different. Crane provides rain data for a variety of rain probabilities, but does not use an availability adjustment factor like the ITU model. Even so, experience shows that the models generally produce results that are in reasonably close agreement.

The process of determining the linear regression coefficients for the specific attenuation due to rain is the same as for the ITU. In fact, the same coefficients are used. The difference is that the specific rain rate for the desired availability must be used. The rain rates versus probability are given in Table 10A.4 for each of the Crane rain regions.

For ground-to-space links, the height of the rain cell must be taken into account, just as it is with the ITU model. Crane uses rain cell height data that are a function of latitude and probability [12]. As seen in Table 11.2, for 30 degrees latitude, the rain heights are

$$h_{001} = 5.35 \text{ km} \qquad \text{for } p = 0.001\%$$

$$h_1 = 3.94 \text{ km} \qquad \text{for } p = 1\%$$

Crane gives a procedure for doing a logarithmic interpolation over availability [13]:

$$H_R(p) = h_1 + \frac{h_{001} - h_1}{\ln(0.001)}\ln(p) \tag{11.26}$$

Figure 11.6 shows the interpolated rain cell height for several different latitudes. For other latitudes, values from Table 11.2 should be used in (11.26).

TABLE 11.2   Crane Model Rain Cell Heights for Different Latitudes

Latitude, North or South (deg)	Rain Height (km)	
	0.001%	1%
≤2	5.3	4.6
4	5.31	4.6
6	5.32	4.6
8	5.34	4.59
10	5.37	4.58
12	5.4	4.56
14	5.44	4.53
16	5.47	4.5
18	5.49	4.47
20	5.5	4.42
22	5.5	4.37
24	5.49	4.3
26	5.46	4.2
28	5.41	4.09
30	5.35	3.94
32	5.28	3.76
34	5.19	3.55
36	5.1	3.31
38	5	3.05
40	4.89	2.74
42	4.77	2.45
44	4.64	2.16
46	4.5	1.89
48	4.35	1.63
50	4.2	1.4
52	4.04	1.19
54	3.86	1
56	3.69	0.81
58	3.5	0.67
60	3.31	0.51
62	3.14	0.5
64	2.96	0.5
66	2.8	0.5
68	2.62	0.5
≥70	2.46	0.5

Source: Table 4.1 from Ref. 12, courtesy of John Wiley & Sons.

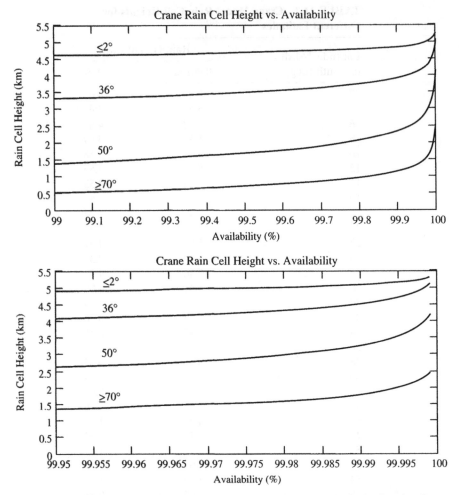

Figure 11.6 Rain cell height versus availability at 30 degrees latitude, for the Crane global model.

The approximate path length in rain is then

$$d \approx \frac{2(H_R - H_S)}{\tan(\zeta) + \sqrt{\tan^2(\zeta) + 2(H_R - H_S)/R_e}} \qquad (11.27)$$

where

H_R is the effective rain cell height
H_S is the effective station height
ζ is the elevation angle
R_e is the 4/3 earth radius, 8500 km

The expression for terrestrial path rain attenuation is employed to determine the horizontal path attenuation.

$$Atten = k \cdot RR^{\alpha} (e^{y \cdot \delta} - 1)/y \quad \text{dB}, \qquad 0 < d < \delta(RR) \text{ km} \qquad (11.28)$$

and

$$Atten = k \cdot RR^{\alpha} \left[\frac{e^{y \cdot \delta(RR)} - 1}{y} + \frac{(e^{z \cdot d} - e^{z \cdot \delta(RR)}) \cdot e^{0.83 - 0.17 \ln(RR)}}{z} \right] \quad \text{dB},$$
$$\delta(RR) < d < 22.5 \text{ km} \qquad (11.29)$$

where $\delta(RR)$ is a function of the rain rate,

$$\delta(RR) = 3.8 - 0.6 \ln(RR) \qquad \text{km} \qquad (11.30)$$

d is link distance in km,

$$y = \alpha \cdot \left[\frac{0.83 - 0.17 \ln(RR)}{\delta(RR)} + 0.26 - 0.03 \ln(RR) \right] \qquad (11.31)$$

and

$$z = \alpha \cdot (0.026 - 0.03 \ln(RR)) \qquad (11.32)$$

The total slant-path attenuation due to rain is then given by

$$A_S = \frac{L}{d} Atten \qquad (11.33)$$

where

$$L = \sqrt{(H_s - R_e)^2 \sin^2(\zeta) + 2 R_e (H_R - H_S) + H_R^2 - H_S^2} - (H_S + R_e) \sin(\zeta) \qquad (11.34)$$

The overall attenuation on an earth–space path ends up being a function of the rain rate (availability), frequency and elevation angle, just as it does with the ITU model. Figure 11.7 shows plots of the expected rain attenuation on an earth–space path versus availability, for several elevation angles and operating frequencies using the Crane model in rain region E.

When applying the Crane global model to slant paths with elevation angles less than about 10 degrees, depending upon the availability and the rain cell height, it is possible to compute a slant-path distance, d, that is greater than 22.5 km. Since this is beyond the valid range of the Crane model, the values for $Atten(RR,d)$ and hence A_s are invalid. This can lead to bogus attenuation

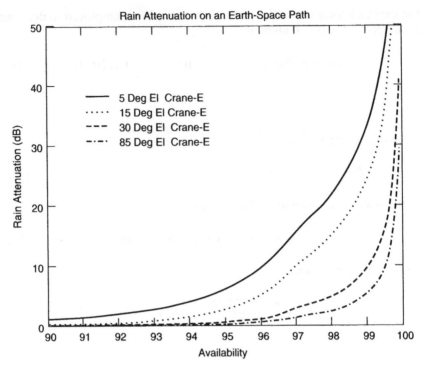

Figure 11.7   Rain attenuation values versus availability for different elevation angles at 30 GHz using the Crane model in rain region E, at 30 degrees latitude.

estimates that can increase as the availability decreases. The author has found that a viable work around for this circumstance is to replace *Atten(RR,d)* by *Atten(RR,22.5)* in the expression for A_s whenever $d > 22.5$ km.

Example 11.4.   A ground station must communicate with a high-flying aircraft with 99.9% rain availability. The aircraft is flying at 50,000 ft and may appear at elevation angles as low as 15 degrees. What fade margin should be used if the ground station is located in Southern California and the frequency of operation is 10 GHz?

The aircraft altitude ensures that it will be flying above any heavy rainfall, so this problem may be treated as a satellite rain attenuation problem. Southern California is in Crane rain region F and it is assumed that the latitude is 32 degrees, that the earth station altitude is near sea level, and that circular polarization is used.

The regression coefficients for circular polarization at 10 GHz are readily found to be

$$k = 0.009485, \qquad \alpha = 1.270$$

using values from Table 10A.2 and $\tau = 45$ degrees for circular polarization. The 99.9% rain rate is found in Table 10A.4:

$$RR = 5.3 \text{ mm/h}$$

The rain height is found by interpolating values from Table 11.2 for 32 degrees:

$$H_R(0.1\%) = 3.76 + \frac{5.28 - 3.76}{\ln(0.001)} \ln(0.1) = 4.27 \text{ km}$$

The path distance in rain is given by

$$d \approx \frac{2(4.27 - 0)}{\tan(15) + \sqrt{\tan^2(15) + 2(4.27 - 0)/8500}} = 15.87 \text{ km}$$

The next step is to compute $\delta(RR)$:

$$\delta(5.3) = 2.799 \text{ km}$$

Since $d > \delta(RR)$, the horizontal path attenuation is given by

$$Atten = k \cdot RR^\alpha \left[\frac{e^{y \cdot \delta(RR)} - 1}{y} + \frac{\left(e^{z \cdot d} - e^{z \cdot \delta(RR)}\right) \cdot e^{0.83 - 0.17\ln(RR)}}{z} \right] \quad \text{dB}$$

where

$$z = -0.0305$$

and

$$y = 0.2175$$

The resulting attenuation is then

$$Atten = 2.45 \text{ dB}$$

The total slant-path attenuation due to rain for 99.9% availability is then given by

$$A_s = \frac{L}{d} Atten$$

where

$$L = \sqrt{(H_S + R_e)^2 \sin^2(\zeta) + 2R_e(H_R - H_S) + H_R^2 - H_S^2} - (H_S + R_e)\sin(\zeta)$$

The result is

$$A_S = 2.46 \text{ dB}$$

It is clear that the dry climate in California combined with a lower frequency considerably reduces the impact of rain on link operation. □

11.7.3 The DAH Rain Attenuation Model

The DAH rain attenuation model, developed by Dissanayake, Allnut, and Haidara [14, 15] and also sometimes called the USA model, is a modification of the ITU model presented earlier. The essential difference of the DAH model is that it accounts for the fact that rain is nonhomogeneous in both the vertical and horizontal directions. The DAH model can be implemented using the following 10-step procedure [14].

Step 1: Compute the freezing height of the rain (analogous to the rain height for the ITU model) using the station latitude, ϕ.

$$h_{fr} = 5, \qquad 0° < \phi < 23° \tag{11.35a}$$

$$h_{fr} = 5.0 - 0.075(\phi - 23), \qquad \phi \geq 23° \tag{11.35b}$$

Step 2: Determine the slant path below the freezing rain using the same procedure as the ITU model.
For $\theta < 5$ degrees,

$$L_s = \frac{2(h_{fr} - h_s)}{\left(\sin^2(\theta) + \dfrac{2(h_{fr} - h_s)}{R_e}\right)^{1/2} + \sin(\theta)} \quad \text{km} \tag{11.36a}$$

Otherwise,

$$L_s = \frac{(h_R - h_s)}{\sin(\theta)} \quad \text{km} \tag{11.36b}$$

where

θ is the elevation angle
h_s is the station height ASL in km

Step 3: Determine the horizontal projection of the slant-path length, just as was done for the ITU model.

$$L_G = L_s \cos(\theta) \tag{11.37}$$

Step 4: Determine the specific attenuation due to rain for 0.01% availability and the desired frequency and polarization, the same as for the ITU model.

$$\gamma_R = k(R_{0.01})^\alpha \quad \text{dB/km} \tag{11.38}$$

Step 5: Compute the horizontal path adjustment factor for 0.01% of the time using the same expression as was used for the ITU model.

$$rh_{0.01} = \frac{1}{1 + 0.78\sqrt{\dfrac{L_G \gamma_R}{f}} - 0.38(1 - e^{-2L_G})} \tag{11.39}$$

where L_G is in km and f is in GHz.

Step 6: Calculate the adjusted rainy path length.

$$L_r = \frac{L_g rh_{0.01}}{\cos(\theta)} \quad \text{for } \zeta > \theta \tag{11.40a}$$

$$L_r = \frac{h_{fr} - h_s}{\sin(\theta)} \quad \text{for } \zeta \le \theta \tag{11.40b}$$

where

$$\zeta = \tan^{-1}\left(\frac{h_{fr} - h_s}{L_g rh_{0.01}}\right) \tag{11.41}$$

Step 7: Compute the vertical adjustment factor analogous to the ITU model.

$$rv_{0.01} = \frac{1}{1 + \sqrt{\sin(\theta)}\left(31(1 - e^{-(\theta/(1+\chi))})\dfrac{\sqrt{L_R \gamma_R}}{f^2} - 0.45\right)} \tag{11.42}$$

where

$$\chi = 36 - |\phi| \quad \text{for } |\phi| < 36° \tag{11.43a}$$

$$\chi = 0 \quad \text{for } |\phi| \ge 36° \tag{11.43b}$$

Step 8: Find the effective path length through rain.

$$L_e = L_r r v_{0.01} \quad \text{km} \tag{11.44}$$

Step 9: The attenuation exceeded 0.01% of the time is then given by

$$A_{0.01} = \gamma_R L_e \quad \text{dB} \tag{11.45}$$

Step 10: To find the attenuation for other percentages between 0.001 and 10%, the same adjustment procedure is used as step 10 of the ITU model.

$$A_p = A_{0.01} \left(\frac{p}{0.01} \right)^{-(0.655 + 0.033 \ln(p) - 0.045 \ln(A_{0.01}) - \beta(1-p)\sin(\theta))} \tag{11.46}$$

where

$$\text{If } p \geq 1\% \text{ or if } |\phi| \geq 36°, \quad \beta = 0 \tag{11.47a}$$

$$\text{If } p < 1\% \text{ and } |\phi| < 36° \text{ and } \theta > 25°,$$
$$\beta = -0.005(|\phi| - 36°) \tag{11.47b}$$

Otherwise,

$$\beta = -0.005(|\phi| - 36°) + 1.8 - 4.25\sin(\theta) \tag{11.47c}$$

In their joint paper, Dissanayake et al. [14] also presented a combined effects model that provides a means modeling all of the link impairments together. The model is based on fitting empirical data and addresses the interdependence between the various link impairments. □

Rather than providing the substantial link margins that are required at millimeter-wave frequencies, another approach to mitigating the impact of rain fades is path diversity, where two antennas and transceiver pairs, separated by a few kilometers or more, are used to perform communications. This is valuable for hardware redundancy (increased reliability), limiting the effect of rain fades and mitigating the effects of ionospheric and tropospheric fades and multipath. This technique exploits the localized nature of very heavy rainfall. At frequencies and availabilities where the fade margins are large, using path diversity may be more cost effective than designing in the required margin.

11.8 ANTENNA CONSIDERATIONS

Antenna gain is a function of the aperture size and the design beamwidth. For point-to-point GEO transponders, narrow-beam high-gain antennas may be used to provide increased gain, thereby requiring less transmit power. This has the added benefit of reducing the susceptibility to interference (spatial filtering), but the antennas then require careful pointing. Broad coverage GEO satellites such as Direct TV will use a broad-beam antenna for area coverage.

For non-GEO satellites, wider beamwidths are used to increased coverage area at the expense of reduced gain. This approach eases the tracking and pointing requirements. MEO and LEO systems may use tracking dish antennas on the ground, but others use broad-beam antennas. GPS satellites actually use a shaped beam to equalize signal strength across the entire coverage area. Direct TV does a similar thing to adjust for the longer path length to northern customers. GPS receivers use a hemispherical antenna pattern to enable it to see all satellites that are above the horizon at a given time.

In addition to the attenuation caused by rain, numerous experiments [3, 16, 17] have shown that water accumulation on the radome or on the reflector and feed lens of the ground-based antenna can be a significant source of signal loss. The amount of loss depends on several factors, including radome or lens material, frequency, orientation, and temperature. Even a low-loss radome can become lossy when wet. The amount of attenuation depends on whether the water is in droplets or as a sheet of water. Radomes and feed lenses that do not have hydrophobic coatings can actually absorb water into the surface, producing attenuation. Icing of a radome can also be a significant source of attenuation, particularly when the ice is melting and forms a water layer on each side of the ice sheet.

The ACTS experimental results for wetting of the antenna feed and reflector indicate losses between 1 and 6 dB, with the greater losses being at the higher frequency (27 GHz) and higher rain rates [17]. The impact of water on the feed lens is greater than that of the water on the reflector according to Ref. 16, whereas Ref. 3 concluded that water on the reflector is the largest contributor. The feeds were covered by plastic lenses, not unlike the material from which regular radomes are made. Crane [17] indicates that the loss due to a wet feed was very low initially due to the hydrophobic coating of the lens, but after a few months the loss became substantial when the lens got wet.

It is also interesting that significant wet antenna attenuation was observed in the ACTS experiment during a heavy fog, with no rain [17]. In general, the heavier rain will have the heavier radome or antenna losses because of the water sheeting across the protective surfaces.

The safest approach to modeling this effect at the frequency of interest is to perform an experiment on a clear day with different misting rates. This can be performed at the desired frequency and orientation. For the most robust results, an aged (weathered) feed lens or radome should be used to account for the hydroscopic property of weathered plastic.

11.9 NOISE TEMPERATURE

Most terrestrial communication systems have noise figures in the 4- to 10-dB range due to internally generated noise and assume an input noise temperature (antenna temperature) of $T_0 = 290\,K$. This makes sense since the noise temperature of the surrounding environment is nominally 290 K, so having an extremely low noise receiver is of limited value, as will be seen in the next section. Antenna noise is due to any noise sources, including absorptive/resistive losses that are in the field of view of the antenna. Ground-based satellite receivers, on the other hand, look skyward, seeing a relatively low noise temperature and can therefore take advantage of receivers with lower noise figures.

For sophisticated satellite systems, external noise is often the limiting factor on performance. The satellite transmit power is extremely limited, however, so the ground-based receiver is usually required to be very sensitive or to have a very high gain antenna. Since the ground-based receiver is looking skyward, it is subject to atmospheric, ionospheric, and galactic noise as well as to the internal receiver noise floor. The receiver on the satellite is not required to be as sensitive since, in general, the transmit power from earth is not as limited. The noise floor of the satellite-based receiver is limited by the earth background ($T = 290\,K$) and is also subject to noise from the atmosphere.

The antenna noise temperature is set by the noise that the antenna "sees" or receives. Rain and the earth are generally taken to have $T = 290\,K$. For extreme conditions, the value of 323 K is sometimes used, which is 50°C. Sky noise is a function of frequency and of elevation angle due to atmospheric absorption. The net antenna temperature is a function of sky noise temperature (due to galactic noise and atmospheric loss) and ground temperature as well as of the antenna gain, antenna beamwidth (unit solid angle), and loss within the antenna. To accurately compute the antenna temperature, the analyst must integrate the product of the antenna gain and the apparent temperatures of the sky, earth, and sun or moon over the entire sphere of reception. This requires complete knowledge of the antenna radiation pattern and the temperature of the surrounding earth. The antenna temperature can be approximated by [18]

$$T_a = a_1 T_{sky} + a_2 T_g + a_3 T_{sun} \tag{11.48}$$

where

$$a_1 = \frac{1}{4\pi}(G_{sky}\Omega_{sky} + \rho^2 G_g \Omega_g)$$

$$a_2 = \frac{\Omega_g}{4\pi} G_g (1-\rho)^2$$

$$a_3 = \rho \frac{\Omega_s}{4\pi} \frac{G_s}{A_r}$$

$$\Omega_{SKY} \sim \theta_{AZ}\theta_{EL}$$

ρ is the (voltage) reflection factor of the ground.

T_s is the apparent sun temperature (36,500 K).

p is the polarization loss factor for solar radiation since solar radiation is unpolarized (0.5).

The Ω terms are the unit solid angles for the sky, earth, and sun.

The G terms are the average gains over the unit solid angles given by the respective Ω values.

A_r is the rain attenuation (if present).

Figure 11.8 is a plot of sky temperature versus frequency. As expected, the lower elevation angles that pass through more of the atmosphere show the higher temperatures.

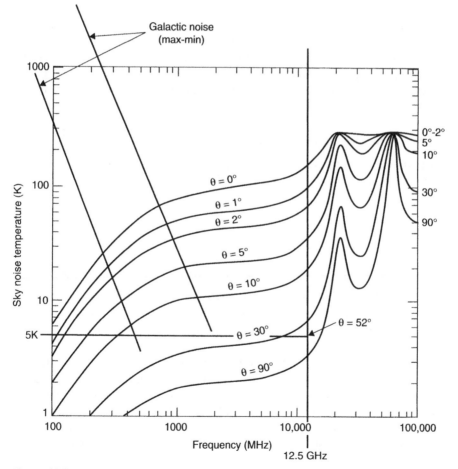

Figure 11.8 Sky noise temperature versus frequency for different elevation angles. (Figure 71.3 from Ref. 18, courtesy of CRC Press.)

While the relationship between equivalent noise temperature and the noise figure was established in Chapter 3, the noise figure is not generally used for satellite work. This is for two reasons. First, the definition of noise figure assumes an input temperature of 290 K for the source. In the case of satellite systems, the antenna temperature may not be 290 and thus it does not have a standard temperature input. The other reason is that satellite systems typically use very high quality front-end components, which would imply noise figures very close to 1 (0 dB). The effective noise temperature is the additional temperature over the standard noise temperature. For satellite systems, a system temperature is used instead, which represents the total temperature of the system, so that the system noise power spectral density is given by

$$N_0 = kT_{sys} \tag{11.49}$$

where k is Boltzmann's constant.

An important parameter in the characterization of satellite receivers is the ratio of the antenna gain to the system noise temperature, G/T. This ratio is sometimes called the sensitivity or the system figure of merit. The use of this parameter emphasizes the fact that increasing the receive antenna gain or decreasing the system noise temperature will improve receiver performance.

11.9.1   The Hot-Pad Formula

An attenuator in the signal path increases the effective noise input temperature according to the hot-pad formula [19]

$$\boxed{T_N = (T_{in} + (L-1)T)/L} \tag{11.50}$$

where

T_N is the resulting temperature at the antenna input (attenuator output)
T_{in} is the nonattenuated noise temperature at the input (T_{SKY})
T is the temperature of the attenuator, usually: $T_0 = 290$ K, sometimes 323 K is used for warm climates and extreme conditions

The validity of the hot-pad formula can be heuristically demonstrated as follows. Consider a perfect antenna pointing toward the sky. If the frequency is 10 GHz and the elevation angle is 10 degrees, then $T_{sky} \sim 15$ K, which becomes T_{in} in the hot-pad formula. Now consider the effect of additional attenuation, perhaps tree foliage. If there is no attenuation, then $L = 1$ and the new sky temperature should still be equal to T_{sky}. If the attenuation were infinite, then the antenna would simply see thermal noise and the new sky temperature would be T_0, which the hot-pad formula indicates. For other cases where the loss is finite, the resulting sky temperature will fall between 15 K and 290 K, depending upon the actual value of the loss. For a 10-dB attenuation, the new T_{sky} becomes 262.5 K, whereas for 1-dB attenuation, T_{sky} is 71.6 K.

By applying the hot-pad formula, it is apparent that the effect of a loss in front of the receiver LNA can be both a signal reduction and an increase in the noise level. The hot-pad formula can be applied to any passive attenuation that is present between the noise source and the receiver. In the case of rain attenuation, the loss is not strictly between the atmospheric noise and the receiver; however, the hot-pad formula is still often used for this application.

Example 11.5. Consider a system with an antenna temperature of 20 K, an antenna gain of 30 dB, and a maximum expected rain fade of 6 dB. If the receiver front end has an effective noise temperature of 50 K, what is the overall system noise temperature in the clear and during a rain fade?

The clear-air system noise temperature is simply

$$T_{sys} = T_a + T_e = 70 \text{ K}$$

During a rain fade, the hot-pad formula must be applied to determine the adjusted noise temperature.

$$T_N = (T_{in} + (L-1)T)/L$$

where T_{in} is T_a, L is 4 (6 dB), and T is the temperature of the fade, in this case rain ($T_0 = 290$ K). Thus, $T_N = 222.5$ K and $T_{sys} = 272.5$ K.

The noise in the system increases by 272.5/70, or 5.9 dB, due to the absorptive loss of the rain. At the same time, the signal decreases by L or 6 dB, so the total reduction in SNR is 9.9 dB! The system figure of merit is

$$G/T = 1000/70, \quad \text{or } 14.3 \text{ dB}$$

The above analysis can be repeated for the case where the receiver temperature is 300 K instead of 50 K. The antenna gain is adjusted to keep the clear-air SNR the same (i.e., maintain the same clear-air G/T, thus the antenna gain is higher).

$$G = 1000(320/70), \quad \text{or } 36.6 \text{ dB}$$

The clear-air system noise temperature is simply

$$T_{sys} = T_a + T_e = 320 \text{ K}$$

During a rain fade, the hot-pad formula is applied;

$$T_N = (T_{in} + (L-1)T)/L$$

where T_{in} is T_a, L is 4 (6 dB) and T is the temperature of the fade, in this case rain ($T_0 = 290$ K). Thus, $T_N = 222.5$ K and $T_{sys} = 522.5$ K. So the noise in the system increases by 522.5/320, or 2.1 dB. The signal still decreases by L or 6 dB, and the total reduction in SNR is 8.1 dB. The system figure of merit is

$$G/T = 1000(320/70)/320$$

or still 11.5 dB. This shows that in some cases, trading noise temperature for antenna gain is worthwhile. Put another way, if there are significant absorptive losses in the link, resources are best used to increase antenna gain rather than reducing the system noise temperature below ambient.

The application of the hot-pad formula to rain or other atmospheric effects actually applies only to the sky-noise temperature and not the total antenna temperature. The portion of the antenna temperature that is due to blackbody radiation from the earth will not be affected by the rain attenuation in the same way (although the earth may be cooled by the rain at the same time), so in systems where the antenna temperature exceeds the sky-noise temperature significantly, the application of the hot-pad formula must be re-worked to only modify the sky noise and then add the other contributions to antenna temperature afterwards.

11.9.2 Noise Due to Rains

The effect of rain attenuation will be to reduce the signal and to cause an increase in the received noise level. Since the rain is not actually between the atmospheric noise and the antenna (they are intermingled), the hot-pad formula is not the most precise way to account for the temperature increase. Instead, a rain temperature can be computed and added to the system temperature. The rain temperature is given by [20]

$$T_r = T_m \left(1 - 10^{\frac{-A_r(dB)}{10}} \right) \quad \text{K} \tag{11.51}$$

where T_m is the mean path temperature. The mean path temperature can be estimated from

$$T_m = 1.12T_s - 50 \text{ K} \tag{11.52}$$

where T_s is the surface temperature of the surrounding area. Generally, simply using a value of $T_m = 273$ K provides good results.

Example 11.6. Consider the system from Example 11.5, with an antenna temperature of 20 K, an antenna gain of 30 dB, and a maximum expected rain fade of 6 dB. If the receiver front end has an effective noise temperature of 50 K, what is the overall system noise temperature in the clear and during a rain fade?

The clear-air system noise temperature is simply

$$T_{sys} = T_a + T_e = 70 \text{ K}$$

During a rain fade, the rain temperature must be added to the sky or antenna temperature to determine the adjusted noise temperature.

$$T_r = T_m \left(1 - 10^{\frac{-A_r(dB)}{10}} \right) \quad \text{K}$$

where T_m is taken to be 273 and A_r is 6 dB. Thus, $T_r = 204.4$ K and T_{sys_rain} becomes 274.4 K.

The noise in the system increases by 274.4/70, or 5.9 dB due to the absorptive loss of the rain. At the same time, the signal decreases by A_r or 6 dB, so the total reduction in SNR is again 11.9 dB. So, for this example, the result is virtually identical (particularly when expressed in dB) to applying the hot-pad formula with a rain temperature of 290 K. ☐

11.9.3 Sun Outages

The sun represents a very high temperature noise source, which subtends approximately 0.5 degrees. When the sun passes through the main beam of a high-gain antenna connected to a low-noise receiver, the effect on the received signal can be dramatic. This should be apparent from the preceding expression for the antenna temperature. If sufficient margin is not available, communications will be temporarily lost. This is called a sun outage (or sun transit outage) and occasionally occurs on satellite feeds. The phenomenon is readily predictable, and remedies include using alternate satellite paths (diversity), using very large link margins, or tolerating the brief outages. Some communication satellite operating companies provide online calculators to predict sun outages given the latitude and longitude of the planned ground station.

The solar noise level is approximately −183 dBW/Hz (compared to −204 dBW/Hz for thermal noise). The solar noise must be integrated over the antenna gain pattern. If the beam width of the antenna is less than 0.5 degrees, it is possible for the sun to fill the entire field of view, resulting in a noise floor of −183 dBW/Hz ($T \sim 36{,}500$ K). This represents a 21-dB increase in the noise level.

11.10 SUMMARY

In this chapter the different types of satellite and satellite orbits are briefly examined, followed by determination of the slant range to the satellite. Next, the impairments to satellite communication link were discussed. Computation of free-space loss to a satellite hinges on proper computation of the distance to the satellite, which is a function of the satellite location and the ground station location. The elevation angle is also determined from the relative geometry and is used in computation of atmospheric and rain attenuation. Atmospheric attenuation is a function of the local climate, the ground station elevation, and, as mentioned, the elevation angle to the satellite. Some of these are similar to terrestrial communications, and some, like ionospheric scintillation, are unique to satellite communication.

Ionospheric effects include scintillation, Faradary rotation, time delay, and time dispersion. The effects are reduced above 1 GHz and virtually nonexistent above 10 GHz. As frequencies increase, the effect of rain on a satellite communication link becomes significant. The usual approach to addressing rain loss is a probabilistic method, where an availability is specified and the required fade margin to achieve that availability is determined and incorporated into the link budget, just as it is for terrestrial links. In this chapter, application of both the ITU model and the Crane global model to satellite links is discussed.

The subject of water on a ground station antenna or radome was also discussed. Losses from wet feeds, reflectors, and/or radomes can range from 1 to 6 dB, but definitive results are not widely available. In addition to the internally generated thermal noise in the receiver, sensitive satellite receivers must also account for external noise or so-called antenna noise. The sources may be galactic or due to resistive losses in the RF path, including radome losses, rain, atmospheric attenuation, and others. The effect of such external noise can be devastating if not planned for. The hot-pad formula provides the means of predicting the impact of such external noise sources. The chapter concludes with a brief discussion of sun-transit outages, which can temporarily blind a ground-based satellite receiver when the sun passes through the field of view of the receive antenna.

REFERENCES

1. ITU Recommendations, *Propagation data and prediction methods required for the design of Earth-space telecommunication systems*, ITU-R P.618-7, 2001.

2. T. Pratt, C. Bostian, and J. Allnutt, *Satellite Communications*, 2nd ed., Wiley, Hoboken, NJ, 2003, p. 19.

3. R. K. Crane, X. Wang, D. B. Westenhaver, and W. J. Vogel, ACTS Propagation Experiment: Experiment Design, Calibration, and Data Preparation and Archival, Special Issue on Ka-Band Propagation Effects on Earth-Satellite Links, *Proceedings of the IEEE*, Vol. 85, No. 6, June 1997, pp. 863–878.

4. T. Pratt, C. Bostian, and J. Allnutt, *Satellite Communications*, 2nd ed., Wiley, Hoboken, NJ, 2003, pp. 32–34.

5. M. I. Skolnik, *Introduction to Radar Systems*, 3rd ed., McGraw-Hill, New York, 2001, p. 500.

6. ITU Recommendations, *Attenuation by atmospheric gasses*, ITU-R P.676-5, Geneva, 2001.

7. ITU Recommendations, *Characteristics of precipitation for propagation modeling*, ITU-R PN.837-1, Geneva, 1994.

8. ITU Recommendations, *Rain height model for prediction methods*, ITU-R P.839-3, Geneva, 2001.

9. ITU Recommendations, *Rain height model for prediction methods*, ITU-R P.839-1, Geneva, 1997.

10. R. K. Crane, Prediction of attenuation by rain, *IEEE Transactions on Communications*, Vol. 28, No. 9, September, 1980, pp. 1717–1733.

11. R. K. Crane, *Electromagnetic Wave Propagation Through Rain*, Wiley, Hoboken, NJ, 1996, pp. 110–119.

12. R. K. Crane, *Electromagnetic Wave Propagation Through Rain*, Wiley, Hoboken, NJ, 1996, p. 150.

13. R. K. Crane, *Electromagnetic Wave Propagation Through Rain*, Wiley, Hoboken, NJ, 1996, p. 149.

14. A. W. Dissanayake, J. E. Allnut, and F. Haidara, A prediction model that combines rain attenuation and other propagation impairments along earth-satellite paths, *IEEE Transactions on Antennas and Propagation*, Vol. 45, No. 10, October 1997, pp. 1546–1558.

15. L. J. Ippolito, *Propagation Effects Handbook for Satellite System Design*, 5th ed., Stanford Telecom, Pasadena, CA, 1999, pp. 2–86.

16. R. J. Acosta, Special Effects: Antenna Wetting, Short Distance Diversity and Depolarization, *Online Journal of Space Communication*, Issue No. 2, Fall 2002.

17. R. K. Crane, Analysis of the effects of water on the ACTS propagation terminal antenna, special issue on Ka-band propagation effects on earth-satellite links, *Proceedings of the IEEE*, Vol. 85, No. 6, June 1997, pp. 954–965.

18. J. D. Gibson, *The Communications Handbook*, CRC Press, Boca Raton, FL, 1997, pp. 970–972.

19. J. D. Gibson, *The Communications Handbook*, CRC Press, Boca Raton, FL, 1997, p. 968.

20. L. J. Ippolito, *Propagation Effects Handbook for Satellite System Design*, 5th ed., Stanford Telecom, Pasadena, CA, 1999, pp. 2–168.

EXERCISES

1. Consider a 12-GHz, geostationary satellite located at 105 degrees west longitude. If an earth station is located at 80 degrees west longitude and 35 degrees north latitude, determine the following;

(a) The central angle to the satellite

(b) The elevation angle to the satellite

(c) The free-space loss to the satellite

2. The Direct TV DBS transponders are located at approximately 101 degrees west longitude in geostationary orbit. Using your latitude and longitude, determine the elevation angle to the satellite(s) and estimate the azimuth angle.

3. Consider a 15-GHz satellite link. If the ground station is located in Denver, CO while the satellite is located in geosynchronous orbit at 98 degrees west longitude, determine the following;

(a) The central angle

(b) The elevation angle to the satellite

(c) The slant range to the satellite and the free-space loss

(d) The expected atmospheric attenuation

(e) The depth of a 99.995% fade using the elevation angle found in part b and assuming that the polarization is vertical. State which rain model you use.

4. Consider a 30-GHz satellite link operating from Atlanta, GA at 28 degrees of elevation, using circular polarization.

(a) Determine the depth of a 99.999% rain fade using the Crane global model.

(b) Find the additional signal-to-noise ratio degradation due to the noise from the rain loss.

(c) Determine the expected atmospheric attenuation.

(d) Find the amount of noise floor degradation (noise increase) due to the atmospheric loss if the antenna temperature is 50 K.

5. Repeat parts a and b of problem 4 using the DAH model.

CHAPTER 12

RF Safety

12.1 INTRODUCTION

Few subjects in electrical engineering evoke as much controversy as the debate over safe levels of exposure to electromagnetic energy. There is considerable misunderstanding about the known effects of electromagnetic energy on humans, and public perception can be an important factor in system deployment. While there is still much to be learned about the effects of prolonged exposure to low-power-density electromagnetic waves, sufficient empirical data on moderate power density levels exist to set reasonably safe exposure limits that balance cost and risk [1]. A set of conservative standards has been developed by the Institute of Electrical and Electronics Engineers (IEEE). The development of the FCC limits that are used in the United States was heavily influenced by these standards. The FCC limits and their guidelines for computations are discussed in this chapter.

Systems used in Europe and many parts of Asia are required to meet the European Telecommunications Standards Institute (ETSI) standards rather than FCC standards. Information about the ETSI standards can be found on their web site, www.etsi.org. The ETSI guidelines [2] use a volume of *Health Physics* and relevant CENELEC, IEC, and ITU standards or recommendations for the specific exposure levels.

Any system that is designed to radiate RF energy should be analyzed and/or tested to verify that the RF exposure of the user and the public is within safe limits. Unsafe levels can be reached due to high transmitter power, high antenna gain, close proximity to the transmitting antenna, or any combination thereof. RF safety is a significant concern for many commercial communication systems due to higher power densities sometimes involved and the potential for public exposure. For consumer products, various potential uses and misuses must be considered when performing an RF safety analysis.

The following terms are used in this chapter:

Introduction to RF Propagation, by John S. Seybold
Copyright © 2005 by John Wiley & Sons, Inc.

ANSI	American National Standards Institute
CENELEC	European Committee for Electrotechnical Standardization
EPA	U.S. Environmental Protection Agency
FCC	U.S. Federal Communications Commission
IEC	International Electrotechnical Commission
MPE	Maximum permissible exposure
NCRP	National Council on Radiation Protection and Measurement
SAR	Specific absorption rate
IEEE	Institute of Electrical and Electronic Engineers
Controlled environment	Where energy levels are known and everyone is aware of the presence of RF
Uncontrolled environment	Where energy levels are not known or where people may not be aware of the presence of RF
Ionizing radiation	Radiation of sufficient energy to strip electrons from molecules, X rays, gamma rays, and so on.
Non-ionizing radiation	Generally applies to electromagnetic waves between RF and light

The American National Standards Institute (ANSI) [3] has adopted the IEEE human safety standards for exposure to electromagnetic radiation. These standards apply to non-ionizing radiation and are set to keep exposure well below thermal-hazard levels. The actual standards are available from the IEEE standards group, ANSI [3], and other sources. The FCC uses a hybrid of these standards and the results of a report by the National Council on Radiation Protection and Measurements (NCRP) [4]. Due to the geometric spreading of electromagnetic waves as they radiate from a source, the strongest fields and highest exposure levels occur in close proximity to the transmitting antenna. When highly directional antennas are used, the safe distance will be a function of the angular location relative to the beam direction, with the sidelobes and backlobes requiring less distance. There is also a potential shock hazard (including possible arcing) at the antenna surface, which cannot be ignored. For the purpose of safety analysis, the standard defines two types of exposure called *controlled* and *uncontrolled* exposure, which will be discussed shortly.

The ANSI/IEEE (and FCC) standards are based on large amounts of scientific data, represent a relatively broad consensus, and are conservative, but they are not absolutes. At the time of this writing, adherence to these standards represents good engineering practice in the United States. The steady improvement of digital modulation methods may well reduce the applications of these standards as lower power devices replace today's technology. Ongoing research may also provide more insight into the effects of electromagnetic radiation on the human body and lead to changes in the standards. Studies to date have failed to provide a conclusive link between athermal effects due to mobile phone use and cancer development or promotion [5].

While there was a heightened concern in the early to mid-1990s as cell phone use grew, that the radiation from cell phone handsets in close proximity to the head was responsible for some cases of brain cancer, the sheer number of mobile phone users guarantees that some victims of cancer will be using phones. Other sources of electromagnetic radiation in the home actually provide greater exposure and are also believed to pose minimal risk. In the ELF range, the magnetic fields from appliances such as hair driers, waterbed heaters, and electric blankets are likely to pose a greater risk than electromagnetic fields at higher frequencies, due to the intensity of the fields and the close proximity of their operation. Even so, the risk is minimal and a policy of "prudent avoidance" (avoiding exposure when the cost or inconvenience of doing so is minimal) is all that is recommended at this time [6].

The FCC emphasizes that these safety limits are exposure limits and not emission limits [1] and that the exposure limits only apply to "locations that are accessible to workers or members of the public." Emissions are regulated separately based on application and licensing. The station operator is responsible for maintaining a safe environment for the public and for workers. The FCC requires station RF safety evaluations, with exemptions for certain low-power applications. Guidelines are provided for estimating field intensity based on antenna geometry and transmit power. These guidelines are relatively easy to apply and tend to overestimate the field intensity. If the guidelines indicate a concern, then a more detailed analysis or testing should be performed. That is, the guidelines provide a quick and dirty analysis that is conservative. If the station does not meet the safety standards based on these calculations, a more detailed (precise) analysis should be performed before altering the station design.

12.2 BIOLOGICAL EFFECTS OF RF EXPOSURE

As indicated earlier, RF radiation is non-ionizing radiation. This is due to the fact that the photonic energy at radio frequencies is insufficient to cause ionization [7]. Figure 12.1 shows the electromagnetic spectrum and indicates the delineation of ionizing radiation. For non-ionizing radiation, tissue heating (thermal effect) is the only verified mechanism for tissue damage. Non-thermal or athermal cell damage and mutation is attributed only to ionizing radiation and has not been associated with non-ionizing radiation. Other possible athermal biological effects of non-ionizing radiation have been postulated, but remain unproven. In OET-56 [8] the authors provide an assessment of the state of research on electromagnetic wave exposure:

> At relatively low levels of exposure to RF radiation, i.e., field intensities lower than those that would produce significant and measurable heating, the evidence for production of harmful biological effects is ambiguous and unproven. Such effects have sometimes been referred to as "non-thermal" effects. Several years

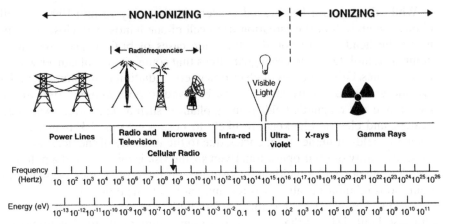

Figure 12.1 Ionizing and non-ionizing radiation as a function of frequency. (Figure 2, page 3 of Ref. 8, courtesy of the FCC.)

ago publications began appearing in the scientific literature, largely overseas, reporting the observation of a wide range of low-level biological effects. However, in many of these cases further experimental research was unable to reproduce these effects. Furthermore, there has been no determination that such effects might indicate a human health hazard, particularly with regard to long-term exposure.

One of the potential effects of electromagnetic wave exposure that is often discussed is cancer. Here again the authors of OET-56 [8] provide some insight.

> Some studies have also examined the possibility of a link between RF and microwave exposure and cancer. Results to date have been inconclusive. While some experimental data have suggested a possible link between exposure and tumor formation in animals exposed under certain specific conditions, the results have not been independently replicated. In fact, other studies have failed to find evidence for a causal link to cancer or any related condition. Further research is underway in several laboratories to help resolve this question.

The authors of OET-56 [8] also indicate that research is continuing and is being monitored.

> In general, while the possibility of "non-thermal" biological effects may exist, whether or not such effects might indicate a human health hazard is not presently known. Further research is needed to determine the generality of such effects and their possible relevance, if any, to human health. In the meantime, standards-setting organizations and government agencies continue to monitor the latest experimental findings to confirm their validity and determine whether alterations in safety limits are needed in order to protect human health.

It is important to appreciate the distinction between a biological *effect* and a biological *hazard*. A small amount of localized tissue heating is a measurable effect, but may not be a hazard. The frequency of the electromagnetic wave and the part of the body exposed are important considerations. The two areas of the body that are most vulnerable to damage from tissue heating are the eyes and the testes as they lack adequate means (blood flow) to rapidly dissipate heat. So while tissue heating is an effect, above a certain level it can become a hazard.

Since tissue heating is the focus, in addition to the electromagnetic field intensity, the duty cycle of the electromagnetic emissions is incorporated into exposure calculations. The result is an averaged power density over time and body surface. Thus the spatial distribution of the time-averaged power density is the unknown that must be determined for a safety analysis.

12.3 FCC GUIDELINES

The current FCC exposure guidelines [1] were adopted in 1996. The guidelines are based on specific absorption rate (SAR) and use a time-averaged whole-body exposure SAR of 4W/kg and include safety margins [9]. The FCC guidelines include "quick and dirty," conservative field calculations to assess compliance with the exposure limits [1]. For the most up-to-date information, visit the FCC Office of Engineering and Technology web site at http://www.fcc.gov/oet/rfsafety. If the FCC computations do not indicate compliance, other, more detailed analysis methods may be used to assess compliance.

Figure 12.2 is a plot of the FCC maximum permissible exposure (MPE) limits in mW/cm² versus frequency in MHz. Both the power density and

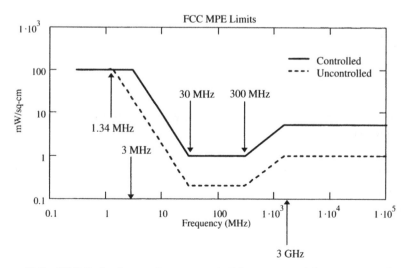

Figure 12.2 FCC limits for maximum permissible exposure (plane-wave equivalent power density).

frequency scales are logarithmic. The lower limits between 30 and 300 MHz reflect the frequency range where the human body most readily absorbs RF energy based on its resonance. Table 12.1 shows the defining expressions for the MPE for controlled exposure, while Table 12.2 shows the expressions for uncontrolled exposure.

In a *controlled environment*, also called occupational/controlled environment, persons must be aware of the potential for exposure and be able to exercise control over their exposure [1]. The required awareness can be achieved by posting warning signs and by providing training programs. The controlled exposure limits apply to the workplace and to amateur radio operators and their immediate families. Examples include workers at transmitter facilities, cellular tower repair and installation workers, and workers at RF test and development facilities.

The exposure limits for *uncontrolled environments* apply in situations where persons may not be aware of their exposure. They also apply in situations where persons are aware of their exposure but cannot do anything to limit it. Examples include nearby cellular towers, wireless office networks, and living or working next door to a radio station (including an amateur radio station). Note that these are examples of applications where the uncontrolled exposure limits apply and are not necessarily hazardous environments.

TABLE 12.1   FCC Maximum Permissible Exposure Limits for Controlled Environment

Frequency (MHz)	E Field (V/m)	H Field (A/m)	Power Density S (mW/cm²)	Averaging Time $|E|^2$, $|H|^2$, or S (min)
0.3–3.0	614	1.63	100[a]	6
3.0–30	1,842/f	4.89/f	$(900/f^2)^a$	6
30–300	61.4	0.163	1.0	6
300–1,500	—	—	f/300	6
1,500–100,000	—	—	5	6

[a] Plane-wave equivalent power density, f is frequency in MHz.

TABLE 12.2   FCC Maximum Permissible Exposure Limits for Uncontrolled Environment

Frequency (MHz)	E Field (V/m)	H Field (A/m)	Power Density S (mW/cm²)	Averaging Time $|E|^2$, $|H|^2$, or S (min)
0.3–1.34	614	1.63	100[a]	30
1.34–30	824/f	2.19/f	$(180/f^2)^a$	30
30–300	27.5	0.073	0.2	30
300–1,500	—	—	f/1,500	30
1,500–100,000	—	—	1	30

[a] Plane-wave equivalent power density, f is frequency in MHz.

Absorption of RF energy is frequency-dependent based on resonance. The (adult) human body absorbs the maximum amount of RF energy at 35 MHz if grounded and 70 MHz if insulated. Parts of the body may resonate at different frequencies and have different sensitivities. Maximum whole-body absorption occurs when the long axis of the body is parallel to the **E** field and is 4/10 of a wavelength [10] (2/10 if grounded since body then acts as a monopole). The frequency of maximum absorption then depends upon the size of the individual, position (arms raised up, squatting down, etc.), and whether or not the individual is grounded.

The maximum permissible exposure (MPE) is defined based on average power level. Thus the signal power (peak envelope power, PEP), transmit duty factor (for push-to-talk or PTT systems), and exposure duration must all be factored in to a computation of exposure. The averaging time is 6 minutes for controlled exposure and 30 minutes for uncontrolled exposure. If the exposure time cannot be controlled, continuous exposure must be assumed. There is also a requirement that the peak power exposure be limited so that an arbitrarily short pulse is not allowed to become arbitrarily powerful. This is addressed by reducing the MPE averaged over any 100 ms by a factor of five for pulses shorter than 100 ms [3].

The MPE limits given in Table 12.1 and Table 12.2 provide electric field, magnetic field, and power densities at frequencies at or below 300 MHz. Above 300 MHz, only power density is specified. When **E** and **H** field limits are given, they must each be met. If both **E** and **H** field limits are met, then the power density limit will also be met. The FCC computations all assume that the **E** and **H** fields are orthogonal (they are not in the near-field), which may not give accurate power density estimates in the near-field of the transmitting antenna.

Maxwell's equations indicate that an electromagnetic wave will have an **E** field and an **H** field that are perpendicular to each other and to the direction of propagation (in the far-field). Thus

$$\mathbf{Z} = \mathbf{E} \times \mathbf{H}$$

yields the power density vector in the direction of propagation. This is a far-field or plane wave computation. The quantity **Z** is called the Poynting vector. Some definitions of the Poynting vector include division by the characteristic impedance so that the units are in power per square meter rather than volt-amps per square meter. It is also noteworthy that some texts use peak values for **E** and **H**, while others use rms values. The use of rms values is fairly standard for safety analysis and is used in this text. It is always important to be aware of the context when using equations for analysis however. If peak values are used, then the equation must also contain a factor of $\frac{1}{2}$ so that the resulting vector is in watts (or normalized watts). Given either **E** or **H** (in the antenna far-field where they are orthogonal), the magnitude of the other can be computed if the characteristic impedance of the medium is known. For free space (air)

$$\eta = 377\ \Omega$$

The characteristic impedance is defined by

$$\eta = \sqrt{\mu/\varepsilon} \qquad (12.1)$$

where μ and ε are the permeability and permittivity of the medium, respectively. The strength of an electric field is expressed in units of volts/meter or equivalent. The strength of a magnetic field is expressed in units of amps/meter or equivalent. In general, electric and magnetic field strengths are given in rms rather than peak values (by convention). Thus the power density can be determined by $|E|^2/\eta$ or $|H|^2\eta$ (analogous to Ohm's Law). Power density is expressed in units of watts/m^2, mW/cm^2, μW/cm^2, or equivalent.

Electromagnetic theory indicates that field strength will decrease with distance. In the far-field, a $1/d^2$ proportionality will apply. In the near-field regions, there can exist areas where the power density is approximately constant with distance (for a short distance) or drops off as $1/d$. Thus as indicated earlier, the fields of interest from a safety standpoint will occur in the vicinity of the transmitting antenna. The purpose of the safety analysis is to determine the region where the RF field strength is low enough that the MPE limits are not exceeded.

12.4 ANTENNA CONSIDERATIONS

For an FCC evaluation, the antenna field of view is divided into four regions, whose definitions differ slightly from those presented in Chapter 3:

The *reactive near-field* region is generally taken to be within one-half wavelength of the antenna surface. Accurate measurement of the fields in this region is difficult due to coupling between the probe and the antenna.

The *near-field* region is

$$\lambda/2 < d < D^2/4\lambda \qquad (12.2)$$

where

d is the distance from the center of radiation of the antenna
D is the largest linear dimension of the antenna
λ is the wavelength

The *transition* region is defined as

$$D^2/4\lambda < d < 0.6D^2/\lambda \qquad (12.3)$$

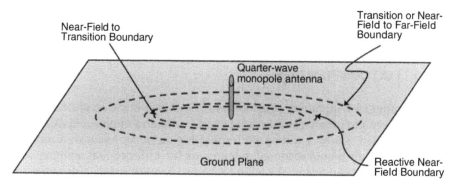

Figure 12.3 Omnidirectional antenna field boundaries for RF safety analysis.

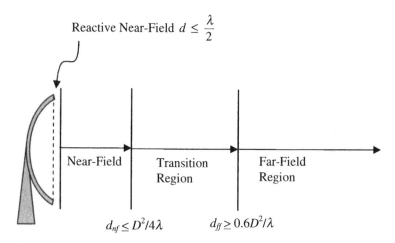

Figure 12.4 Directional antenna field boundaries for RF safety analysis.

The *far-field* region is defined as

$$d > 0.6D^2/\lambda \qquad (12.4)$$

where again D is the largest linear dimension of the antenna.* In the far-field, **E**, **H** and **Z** are all mutually orthogonal and the free space loss equation applies. Figure 12.3 is a conceptual diagram of how the radiation region would appear around a quarter-wave vertical antenna. Figure 12.4 shows the boundaries for a directional (aperture) antenna and includes the boundary definitions. For antennas with dimensions on the order of a wavelength (such as a

---

* This is often taken as $d < 2D/\lambda$ [12]. Since it is not an abrupt boundary, the use of either equation is acceptable for most applications.

monopole) the reactive near-field boundary may fall outside of the transition region, in which case the transition region is not used.

12.5 FCC COMPUTATIONS

The computation equations provided by the FCC are fairly accurate in the far-field and tend to provide conservative (high) estimates of power density in the near-field. The analysis can be performed for the main beam of a direction antenna, for an omnidirectional antenna, or for a directional antenna when outside of the main beam. The main beam and omnidirectional case is treated first, and then the application of the antenna radiation pattern is covered in the following section.

12.5.1 Main Beam and Omnidirectional Antenna Analysis

The estimated power density in the far-field as a function of the distance is given by

$$S = \frac{PG}{4\pi d^2} \quad \text{mW/cm}^2 \tag{12.5}$$

where

P is the transmit power (mW)

G is the antenna gain

d is the distance from the center of radiation of the antenna (cm)

The product of the power and the antenna gain, PG, is the EIRP. The effective isotropic radiated power may be expressed relative to a dipole rather than an isotropic source. In this case it is called *effective radiated power* (ERP) and the value is adjusted by the gain of an ideal dipole in free space.

$$\text{ERP} = \text{EIRP}/1.64 \ (\text{or}, -2.3 \, \text{dB})$$

If the radiator is over a reflective surface, the worst case is a doubling of the electric field intensity. This corresponds to a quadrupling of the power density

$$S = \frac{\text{EIRP}}{\pi d^2} \tag{12.6}$$

The EPA has developed models for ground level power density [1], which suggest a factor of 1.6 rather than 2 for the electric field multiplier over a reflective surface. Applying the EPA result produces a scale factor of 2.56 rather than 4.

$$S = \frac{2.56\text{EIRP}}{4\pi d^2} = \frac{0.64\text{EIRP}}{\pi d^2} \tag{12.7}$$

Example 12.1. Consider a transmitting system shown in Figure 12.5, with

$$f = 100\,\text{MHz}, \qquad \text{ERP} = 10\,\text{kW}$$

The antenna is tower-mounted, with a center of radiation at 50 m and is surrounded by a fence that is 20 m from the base of the tower in all directions. What is the worst-case power density 2 m above the ground at a point 20 m from the base of the tower? Apply the FCC calculations to compute the result.

First, compute the slant distance to the point of interest,

$$d = \left(48^2 + 20^2\right)^{1/2} = 52\,\text{m}$$

Then, assuming that the ground is reflective, the power density is

$$S = \frac{0.64 \cdot 1.64 \cdot \text{ERP}}{\pi d^2} = \frac{1.05 \cdot \text{ERP}}{\pi d^2} = 124\,\mu\text{W}/\text{cm}^2$$

From Table 12.2, the limit for uncontrolled exposure at 100 MHz is $200\,\mu\text{W/cm}^2$. Thus, even this worst-case analysis shows the site to be compliant for uncontrolled exposure, and no further analysis is required. $\square$

12.5.2 Antenna Directivity

If a directional antenna is employed, the antenna gain in the direction of exposure may be much less than the peak gain. In this case, (12.7) can be modified,

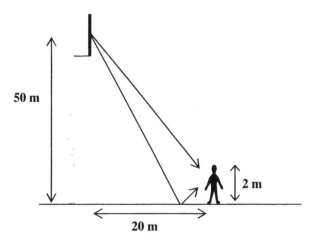

50 m

2 m

20 m

Figure 12.5 RF exposure geometry.

$$S = \frac{0.64 F^2 \text{EIRP}}{\pi d^2} \tag{12.8}$$

where F^2 is the antenna gain reduction in the direction of exposure. For an omnidirectional antenna, $F = 1$. The maximum power density at the surface of an aperture antenna is approximated by

$$S_{surface} = \frac{4P}{A} \tag{12.9}$$

where P is the power to the antenna and A is the physical cross section of the aperture. The maximum value of the near-field, on-axis power density is approximated by

$$S_{nf} = \frac{16 \, \eta P}{\pi D^2} \tag{12.10}$$

and

S_{nf} is the maximum near-field power density
η is the aperture efficiency (0.5–0.75)
P is the power to the antenna
D is the antenna aperture diameter

In the transition region ($D^2/4\lambda < d < 0.6 D^2/\lambda$), the power density can be computed using

$$S_t = \frac{S_{nf} d_{nf}}{d} \tag{12.11}$$

where

S_t is the power density in the transition region
S_{nf} is the maximum near-field power density
d_{nf} is the distance from antenna to the outer edge of the near-field
d is the distance to the point of interest

This is essentially an inverse distance relationship. The far-field computations use the standard free-space loss equation given by (12.5). The free-space loss equation computes the main beam (on-axis) power density. The off-axis power density will, of course, be considerably less.

TABLE 12.3   FCC On-Axis Power Density Equations for Different Antenna Regions

Region	Distance	Field Strength Model
Reactive NF	$0 < d < \lambda/2$	$S_{surface} = \dfrac{4P}{A}$
Near-field	$\lambda/2\pi < d < D^2/(4\lambda)$	$S_{nf} = \dfrac{16\eta P}{\pi D^2}$
Transition	$D^2/(4\lambda) < d < 0.6 D^2/\lambda$	$S_t = \dfrac{S_{nf} d_{nf}}{d}$
Far-field	$0.6 D^2/\lambda < d$	$S = \dfrac{PG}{4\pi d^2}$
Far-field over reflective surface	$0.6 D^2/\lambda < d$	$S = \dfrac{0.64\text{EIRP}}{\pi d^2}$

Example 12.2. Consider a system with the following parameters:

$f = 40\,\text{GHz}$

$P = 20\,\text{dBm}$

30-cm-diameter dish antenna ($G = 40\,\text{dB}$ if $\eta = 0.6$)

Perform a station evaluation on this system. The power density at the antenna surface is

$$S_{surface} = \frac{4P}{\pi D^2/4} = 566\,\mu\text{W}/\text{cm}^2$$

The maximum value of the near-field power density is

$$S_{nf} = \frac{16\eta P}{\pi D^2} = 339.5\,\mu\text{W}/\text{cm}^2$$

The transition region is

$$3.0\,\text{m} < d < 7.2\,\text{m}$$

Within the transition region, the power density is described by

$$S_t = \frac{S_{nf} d_{nf}}{d}$$

where

$$d_{nf} = 3.0 \text{ m}$$

The far-field power density is given by

$$S = \frac{PG}{4\pi d^2}, \qquad d > 7.2 \text{ m}$$

The results are shown in Figure 12.6.

The uncontrolled MPE at 40 GHz is 0.2 mW/cm^2, so this system is compliant for uncontrolled exposure beyond about 5 m. From Table 12.1, the controlled MPE at 40 GHz is 1 mW/cm^2, so the system is compliant everywhere for controlled exposure. Note that all calculations are done in consistent, non-dB units. The results may then be converted to dB (dB μW/cm^2) if desired.

By taking the antenna gain pattern into account, the minimum safe distance would be considerably less than 5 m when outside of the main antenna beam. If the maximum sidelobe level was −15 dB (relative to the peak of the beam), the power density in the far-field would be scaled by

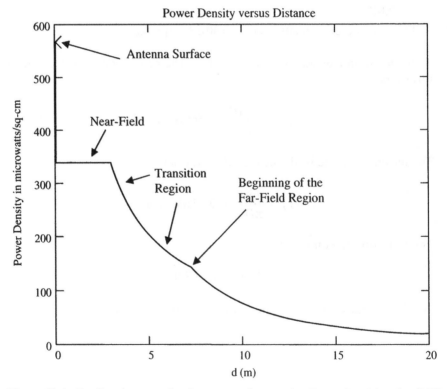

Figure 12.6 Predicted power density versus distance for Example 12.2, using FCC calculations.

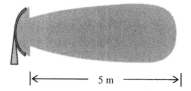

|←——— 5 m ———→|

> Shaded area is restricted for uncontrolled exposure. Controlled exposure is OK at any distance from the antenna, including at the surface

Figure 12.7   Diagram of the directional antenna radiation regions where the FCC MPE limits are exceeded for Example 12.2.

$$G = 0.0316$$

which means that once outside of the antenna main lobe, the system would be safe at any distance up to the antenna surface. Figure 12.7 is a notional diagram of what the hazard area might look like. The height and width of the hazard area would depend upon the elevation and azimuth beamwidths of the antenna and the distance from the antenna center of radiation. □

12.6 STATION EVALUATIONS

All stations are subject to the exposure guidelines, but some installations may be exempted from having to perform routine evaluations depending upon the type of service, power levels involved and the applications. For example, the amateur radio stations that are excluded from the requirement to perform routine evaluations include mobile and portable push-to-talk devices, mobile devices below 1.5 GHz and 1.5 W, mobile devices greater than 1.5 GHz and less than 3 W, and stations that can be shown to be below the MPE limits in all cases. Although some stations are not required to perform routine evaluation of compliance with exposure limits, compliance is still required and an initial station evaluation is in order to verify compliance and to be able to demonstrate compliance if required to do so.

Portable, hand-held consumer products, often with full duty cycle (cellular and cordless phones), typically must use the uncontrolled exposure standard. Devices can be either tested or modeled under typical operating conditions to establish compliance. They should be evaluated based on SAR rather than MPE limits since the close proximity could yield unsafe levels of localized tissue heating while still meeting the whole-body requirements. The MPE limits are based on a whole-body exposure limit of 4 W/kg with safety factors of 10 and 50 for controlled and uncontrolled environments. Thus the actual time-averaged whole-body exposure limits are 0.4 and 0.08 W/kg for controlled and uncontrolled exposure, respectively. For localized exposure, these levels are increased to 1.6 and 8 W/kg as averaged over any 1 g of tissue and averaged over the appropriate time interval [11]. If the SAR limits are met, then it is permissible to exceed MPE limits. In general the MPE is applied for distances greater than 20 cm; otherwise the SAR limit is used. To apply the SAR limits, the designer will usually use a modeling package or construct an instrumented mock-up of the human hand and head and measure the effects.

12.7 SUMMARY

Electromagnetic radiation can be characterized as either ionizing or non-ionizing. Non-ionizing radiation is generally frequencies from zero up to beyond visible light and thus is the type of radiation of interest for RF safety. Presently the only verified effect of non-ionizing radiation is tissue heating. Other effects have been postulated, but remain unproven. The FCC maximum permissible exposure limits were developed based largely on the IEEE/ANSI RF safety standards. The FCC limits are set to levels significantly below where harmful tissue heating occurs. The FCC limits for RF exposure vary with frequency and are different for controlled and uncontrolled environments.

The FCC guidelines include computations that can be used to assess compliance with the maximum permissible exposure limits. These computations are relatively straightforward and provide a conservative estimate of exposure. The computations are only guidelines, however, and if they indicate that the RF exposure is above the limits, a more detailed analysis should be performed. In addition to the frequency and whether individuals are aware of and can exercise any control over their exposure, the FCC limits also take the duty cycle of the transmitted signal and time averaging into account when determining the exposure level.

Mobile and hand-held devices are handled differently as the actual SAR limits (per any 1 g of tissue rather than whole body absorption) must be applied rather than the MPE. This is usually done using measurements or modeling software. The variable nature of hand-held antennas and their position relative to the body make a general MPE difficult to apply and require that the potential for excessive localized exposure and tissue heating be addressed.

REFERENCES

1. R. F. Cleveland, Jr., D. M. Sylvar, and J. L. Ulcek, *Evaluating Compliance with FCC Guidelines for Human Exposure to Radiofrequency Electromagnetic Fields*, OET bulletin 65, Edition 97–01, FCC, August 1997. http://www.fcc.gov/oet/info/documents/bulletins/#65

2. ETSI TR 101 870 V1.1.1 (2001–11), *Fixed Radio Transmitter Site; Exposure to Non-ionising Electromagnetic Fields; Guidelines for Working Conditions*, European Telecommunications Standards Institute.

3. *IEEE Standard for Safety Levels with Respect to Human Exposure to Radio Frequency Electromagnetic Fields, 3 kHz to 300 GHz* (Incorporates IEEE Std C95.1-1991 and IEEE Std C95.1a-1998), ANSI standard C95.1-1999, www.ansi.org

4. E. Hare, *RF Exposure and You*, ARRL, Newington, CT, 1998, p. 4.7.

5. K. R. Foster, and J. E. Moulder, "Are Mobile Phones Safe?", *IEEE Spectrum*, August 2000, pp. 23–28.

6. E. Hare, *RF Exposure and You*, ARRL, Newington, CT, 1998, p. 3.2.

7. R. F. Cleveland, Jr., and J. L. Ulcek, *Questions and Answers about Biological Effects and Potential Hazards of Radio Frequency*, OET bulletin 56, 4th ed., FCC, August 1999, p. 5.

8. R. F. Cleveland, Jr., and J. L. Ulcek, *Questions and Answers about Biological Effects and Potential Hazards of Radio Frequency*, OET bulletin 56, 4th ed., FCC, August 1999, p. 8.

9. E. Hare, *RF Exposure and You*, ARRL, Newington, CT, 1998, p. 4.3.

10. E. Hare, *RF Exposure and You*, ARRL, Newington, CT, 1998, p. 4.2.

11. *IEEE Standard for Safety Levels with Respect to Human Exposure to Radio Frequency Electromagnetic Fields, 3 kHz to 300 GHz* (incorporates IEEE Std C95.1-1991 and IEEE Std C95.1a-1998), ANSI standard C95.1-1999, p. 17. www.ansi.org

12. C. Balanis, *Antenna Theory Analysis and Design*, Wiley, New York, 1997, pp. 32–33.

EXERCISES

1. What is the primary biological effect of RF exposure that is of concern for safety?

2. Determine the reactive near-field, near-field, transition, and far-field boundaries for the following antennas:
   (a) A 200-MHz quarter-wave monopole mounted on a reflective surface
   (b) A 5-GHz dish antenna with a 1-m diameter
   (c) A 7-MHz half-wave horizontal dipole

3. In a controlled exposure environment, the **E** field is measured at 0.05 V/m for 2 min, followed by 0.02 V/m for 4 min and then the cycle is repeated. What is the time-averaged exposure power density in mW/cm^2? The averaging time for controlled exposure is 6 min. Is this permissible exposure?

4. Consider an HF station operating at 30 MHz, with a maximum power of 1 kW supplied to a ground-mounted quarter-wave monopole. If the transmitter duty cycle is 50%, what are the minimum safe distances for controlled and uncontrolled exposure?

5. A satellite ground station, located on the roof of a building uses a 30-dB (directional) antenna pointed upward at 40 degrees to communicate with the satellite using 100 W of continuous transmit power at 4 GHz.

   (a) Is this station governed by controlled or uncontrolled limits? Why?

   (b) What is the minimum safe distance from the antenna?

   (c) What would be the effect of mounting the antenna on a 10-ft tower? That is, how would it affect the answer given in part b?

6. A 2-m (150-MHz) remote weather station link uses a 15-dB yagi antenna and 25 W of transmitter power to communicate with its hub. If the antenna is mounted on a 15-ft tower that is surrounded by a fence, determine the following:

   (a) Is this a controlled or uncontrolled environment?

   (b) What is the minimum safe distance from the antenna?

Review of Probability for Propagation Modeling

This appendix serves as a quick introduction or review of probability theory. While not comprehensive, it does provide the supporting material for the chapters in this book. Heavy reliance on heuristic argument ensures that the mathematically sophisticated reader will be disappointed. Nevertheless, it is felt that for those needing a refresher or a nuts-and-bolts primer, this is the most efficient approach. The author offers his apologies to the purists.

The entire field of probability is founded on the three basic axioms of probability:

$0 \le P(A) \le 1$ The probability of event A occurring is between zero and one.

$P(\rho) = 1$ If ρ is a certain event, then the probability of its occurrence is one.

$P(A + B) = P(A) + P(B)$ If A and B are mutually exclusive events, then the probability of both occurring is equal to the sum of their individual probability of occurrence.

Thus probabilities are properly expressed as real numbers between zero and one inclusive, although sometimes they are written as percentages. At this point it is also helpful to define the concept of relative frequency [1–3]. While not a fundamental axiom of field of probability, it is an aid in relating intuition to probability. Relative frequency is the intuitive concept that the ratio of the number of times that an event occurs to the number of times that the experiment is run (number of trials) will approach the probability of the event as the number of trials gets large.

$$P(A) = \lim_{n \to \infty} \frac{n_A}{n}$$

For the applications in this text, continuous probability is of more interest than discrete, so discrete probability is not addressed other than its use for illustration in some of the examples.

The following definitions are used throughout this appendix:

x is the random variable under consideration.

X is a specific value of x.

A *random variable* is a function that maps the outcome in a sample space to the set of real numbers (i.e., an example of a discrete random variable is a coin toss, where heads $= 1$ and tails $= -1$).

The probability distribution function is defined as

$$F(x) = P(X \le x)$$

For continuous probability, the probability density function (pdf) is defined as the first derivative of the probability distribution function (sometimes called the cumulative distribution function). It should be noted that continuous in this context means that the sample space is defined on a continuum and not that it must be free of discontinuities. It is possible for the probability distribution function to have finite discontinuities, and the corresponding density function will then have impulse functions.

The probability that x is between two values, a and b, is the integral of the pdf from a to b.

$$P(a < x < b) = \int_a^b f(x)\, dx$$

This is illustrated in Figure A.1. Because of the definition of the probability density function and the fact that the random variable is defined on a continuum, the probability of any particular value of x is equal to zero. By the second axiom of probability, the probability of all possible values of x must be unity, so

$$\int_{-\infty}^{\infty} f(x)\, dx = 1$$

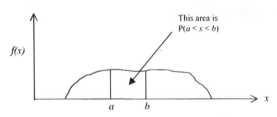

Figure A.1   Notional diagram of an arbitrary probability density function.

One of the pdf's of interest is the uniform pdf, where all values within a particular range are equally likely. An example of a uniform pdf is shown in Figure A.2. Note that the height of the pdf is the reciprocal of the width to ensure that the total probability is one.

The *expected value* or expectation of, x, is defined as

$$E\{x\} = \int_{-\infty}^{\infty} xf(x)\,dx$$

This is the average or mean value of x, also called the first moment and denoted by μ. The mean of the uniform random variable is described by the pdf in Figure A.1, is given by

$$E\{x\} = \int_{-\infty}^{\infty} xf(x)\,dx = \int_{-a/2}^{a/2} x\frac{1}{a}\,dx = 0$$

which is apparent from the pdf.

The second moment of the random variable, x, is defined as

$$E\{x^2\} = \int_{-\infty}^{\infty} x^2 f(x)\,dx$$

This second moment is also called the mean-square value. The second central moment, also called the variance, is the expected value of $(x - \mu)^2$:

$$E\{(x-\mu)^2\} = \int_{-\infty}^{\infty} (x-\mu)^2 f(x)\,dx$$

The variance is equal to the mean-square value minus the square of the mean value. This is shown as follows. The definition of the variance can be written as

$$E\{(x-\mu)^2\} = E\{x^2 - 2\mu x + \mu^2\}$$

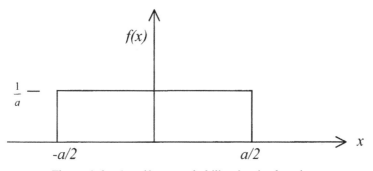

Figure A.2 A uniform probability density function.

By applying the definition of expectation, it is clear that the expectation of the sum of these terms is equal to the sum of the expectations. The expected value of μ^2 is equal to μ^2,* and

$$E\{2\mu x\} = 2\mu E\{x\}$$

Thus the variance of, x, is given by

$$\sigma^2 = E\{x^2\} - \mu^2$$

The standard deviation of a random variable, x, is the square root of the variance of x. It is a metric of the width of the pdf and is denoted by σ. The variance of the uniformly distributed random variable x from Figure A.2 is

$$E\{x^2\} - E^2\{x\} = \int_{-\infty}^{\infty} x^2 f(x)\, dx - \mu^2$$

Since the mean was shown to be zero before, the expression for the variance of this random variable becomes

$$\sigma^2 = \int_{-a/2}^{a/2} x^2 \frac{1}{a}\, dx = \frac{a^2}{12}$$

Example A.1. Given a lot of 100-ohm resistors with a 10% tolerance. The resistors have been tested and any that are out of tolerance have been rejected. What is the pdf, mean, and standard deviation of the resistor population, assuming that all values are equally likely?

From the problem description, the pdf can be readily sketched and will appear similar to the pdf in Figure A.2, where the pdf is centered at 100 ohms instead of zero and $a = 20$. Figure A.3a shows the exact pdf for the resistors.

$$R_{min} = 90, \qquad R_{max} = 110$$

$$f(r) = \begin{cases} 1/20 & \text{for } 90 < r < 110 \\ 0 & \text{everywhere else} \end{cases}$$

The mean value μ equals 100 ohms by inspection. The variance is given by

---

* This uses the fact that the expected value of a constant is that constant. This is readily shown from the definition of expected value and using the fact that $\int_{-\infty}^{\infty} f(x)\, dx = 1$.

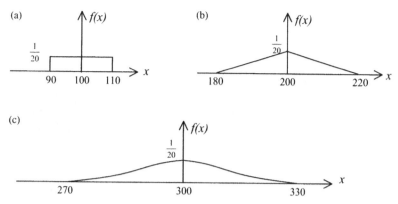

Figure A.3 Pdf's for the sum of uniform random variables. (a) Uniform pdf. (b) Pdf for sum of two uniform random variables. (c) Pdf for sum of three uniform random variables.

$$\sigma^2 = E\{R^2\} - \mu_R^2$$

$$\sigma^2 = \int_{-\infty}^{\infty} R^2 f(R)\, dR - 10{,}000$$

$$\sigma^2 = \int_{90}^{110} 0.05 R^2 dR - 10{,}000$$

$$\sigma^2 = 33.\overline{3}\ \Omega^2$$

which gives a standard deviation of $\sigma = 5.77$ ohms. $\square$

In the preceding example, it is assumed that all permissible values are equally likely. In practice, one might expect the values near the target value (center of the pdf) to occur more frequently (i.e., be more probable). One very common example of such a pdf is the Gaussian or normal pdf.

The Central Limit Theorem of probability states that the pdf of the sum of independent identically distributed (iid) random variables will tend toward a Gaussian pdf as the number of random variables tends to infinity [4–6]. This is an extremely useful theorem that is frequently invoked in probabilistic analyses. It is interesting to note that while the requirement for independence is strict (relaxed only if all of the random variables are already Gaussian), the requirement for being identically distributed is not. As long as the variance of the individual random variables is such that no single or small group of random variables dominates the final distribution, the theorem will apply.

The variance of the resulting Gaussian random variable is given by

$$\sigma^2 = \sum_n \sigma_n^2$$

It is possible to gain an intuitive sense of the Central Limit Theorem by looking at a heuristic demonstration. Suppose two resistors from the previous example are selected at random (independently) and connected in series. This is analogous to adding two independent random variables with the same pdf. The concept of independence will be formally defined shortly. The resulting random variable will have a mean of 2μ, or 200 ohms. From probability theory, the pdf of the sum of two independent random variables is the convolution of the pdf of each of the random variables [7, 8]. Thus the pdf of the series resistance of these two random variables will be triangular as shown in Figure A.3b. Note that the maximum possible value is 220 ohms and the minimum possible value is 180 ohms, which agrees with intuition. The variance of the sum is simply the sum of the individual variances, $\sigma^2 = 66.67$ ohms^2. The triangular shape of the pdf also agrees with intuition. There is only one possible combination of values that gives a series resistance of 220 ohms, but there are many possible combinations that can result in a series resistance of 200 ohms. Thus one would expect higher probability densities near the center of the pdf.

When a third resistor is selected and connected in series, the pdf of the resulting resistance is shown in Figure A.3c. At this point, the tendency toward a Gaussian pdf is clear. In order to be truly Gaussian, the pdf must have infinite tails, so an infinite number of resistors must be used. For a moderate-to-large number of resistors, however, a Gaussian density will be a good approximation.

The Gaussian pdf is described by

$$f(x) = \frac{1}{\sqrt{2\pi}\sigma}e^{-(x-\mu)^2/(2\sigma^2)}$$

and the actual probability that a certain value X is exceeded is given by

$$P(x > X) = \int_X^\infty \frac{1}{\sqrt{2\pi}\sigma}e^{-(x-\mu)^2/(2\sigma^2)}\,dx$$

The preceding integral cannot be evaluated by conventional integration. More advanced techniques like contour integration are required. For engineering applications, the customary procedure is to perform the integration numerically or use a table of standard normal probabilities.

The standard normal pdf has a mean of zero and a standard deviation of unity. Any Gaussian random variable can be converted to an equivalent standard normal random variable using the transformation

$$z = \frac{(x-\mu)}{\sigma}$$

Once the random variable is in standard form, the probability of exceeding any particular value can be looked up in Table A.1. The validity of this trans-

TABLE A.1   Table of Complementary Error Function $Q(z) = \int_{z}^{\infty} \frac{1}{\sqrt{2\pi}} \exp\left(-\frac{u^2}{2}\right) du$

z	0.00	0.01	0.02	0.03	0.04	0.05	0.06	0.07	0.08	0.09
0.0	0.5000	0.4960	0.4920	0.4880	0.4840	0.4801	0.4761	0.4721	0.4681	0.4641
0.1	0.4602	0.4562	0.4522	0.4483	0.4443	0.4404	0.4364	0.4325	0.4286	0.4247
0.2	0.4207	0.4168	0.4129	0.4090	0.4052	0.4013	0.3974	0.3936	0.3897	0.3859
0.3	0.3821	0.3783	0.3745	0.3707	0.3669	0.3632	0.3594	0.3557	0.3520	0.3483
0.4	0.3446	0.3409	0.3372	0.3336	0.3300	0.3264	0.3228	0.3192	0.3156	0.3121
0.5	0.3085	0.3050	0.3015	0.2981	0.2946	0.2912	0.2877	0.2843	0.2810	0.2776
0.6	0.2743	0.2709	0.2676	0.2643	0.2611	0.2578	0.2546	0.2514	0.2483	0.2451
0.7	0.2420	0.2389	0.2358	0.2327	0.2296	0.2266	0.2236	0.2206	0.2168	0.2148
0.8	0.2169	0.2090	0.2061	0.2033	0.2005	0.1977	0.1949	0.1922	0.1894	0.1867
0.9	0.1841	0.1814	0.1788	0.1762	0.1736	0.1711	0.1685	0.1660	0.1635	0.1611
1.0	0.1587	0.1562	0.1539	0.1515	0.1492	0.1469	0.1446	0.1423	0.1401	0.1379
1.1	0.1357	0.1335	0.1314	0.1292	0.1271	0.1251	0.1230	0.1210	0.1190	0.1170
1.2	0.1151	0.1131	0.1112	0.1093	0.1075	0.1056	0.1038	0.1020	0.1003	0.0985
1.3	0.0968	0.0951	0.0934	0.0918	0.0901	0.0885	0.0869	0.0853	0.0838	0.0823
1.4	0.0808	0.0793	0.0778	0.0764	0.0749	0.0735	0.0721	0.0708	0.0694	0.0681
1.5	0.0668	0.0655	0.0643	0.0630	0.0618	0.0606	0.0594	0.0582	0.0571	0.0559
1.6	0.0548	0.0537	0.0526	0.0516	0.0505	0.0495	0.0485	0.0475	0.0465	0.0455
1.7	0.0446	0.0436	0.0427	0.0418	0.0409	0.0401	0.0392	0.0384	0.0375	0.0367
1.8	0.0359	0.0351	0.0344	0.0336	0.0329	0.0322	0.0314	0.0307	0.0301	0.0294
1.9	0.0287	0.0281	0.0274	0.0268	0.0262	0.0256	0.0250	0.0244	0.0239	0.0233
2.0	0.0228	0.0222	0.0217	0.0212	0.0207	0.0202	0.0197	0.0192	0.0188	0.0183
2.1	0.0179	0.0174	0.0170	0.0166	0.0162	0.0158	0.0154	0.0150	0.0146	0.0143
2.2	0.0139	0.0136	0.0132	0.0129	0.0125	0.0122	0.0119	0.0116	0.0113	0.0110
2.3	0.0107	0.0104	0.0102	0.0099	0.0096	0.0094	0.0091	0.0089	0.0087	0.0084
2.4	0.0082	0.0080	0.0078	0.0075	0.0073	0.0071	0.0069	0.0068	0.0066	0.0064
2.5	0.0062	0.0060	0.0059	0.0057	0.0055	0.0054	0.0052	0.0051	0.0049	0.0048
2.6	0.0047	0.0045	0.0044	0.0043	0.0041	0.0040	0.0039	0.0038	0.0037	0.0036
2.7	0.0035	0.0034	0.0033	0.0032	0.0031	0.0030	0.0029	0.0028	0.0027	0.0026
2.8	0.0026	0.0025	0.0024	0.0023	0.0023	0.0022	0.0021	0.0021	0.0020	0.0019
2.9	0.0019	0.0018	0.0018	0.0017	0.0016	0.0016	0.0015	0.0015	0.0014	0.0014
3.0	0.0013	0.0013	0.0013	0.0012	0.0012	0.0011	0.0011	0.0011	0.0010	0.0010
3.1	0.0010	0.0009	0.0009	0.0009	0.0008	0.0008	0.0008	0.0008	0.0007	0.0007
3.2	0.0007	0.0007	0.0006	0.0006	0.0006	0.0006	0.0006	0.0005	0.0005	0.0005
3.3	0.0005	0.0005	0.0005	0.0004	0.0004	0.0004	0.0004	0.0004	0.0004	0.0003
3.4	0.0003	0.0003	0.0003	0.0003	0.0003	0.0003	0.0003	0.0003	0.0003	0.0002

formation and the use of the standard normal (Q-function) table are illustrated in the following example.

Example A.2. Given a random variable with $\mu = 2$ and $\sigma^2 = 9$, what is the probability that the random variable will exceed $X = 6$?

The desired probability is given by

$$P(X > 6) = \int_6^\infty \frac{1}{3\sqrt{2\pi}} e^{-(x-2)^2/18} dx$$

Applying the standard-normal transformation to x yields

$$z = \frac{x-2}{3}$$

and $Z = 1.333$. The sought-after probability becomes

$$P(Z > 1.333) = \int_{1.333}^\infty \frac{1}{\sqrt{2\pi}} e^{-(x)^2/2} dz$$

which is simply $Q(1.333) \approx 0.0912$, from Table A.1. □

Another commonly tabulated function is the error function,

$$\text{erfc}(z) = \frac{2}{\sqrt{\pi}} \int_z^\infty e^{-u^2} du$$

The error function is related to the Q function by the following equation:

$$\text{erfc}(z) = 2Q(\sqrt{2}z)$$

Example A.3. If a batch of 100-ohm resistors has a mean value of 101 ohms and and a standard deviation of 10 ohms, what is the probability of selecting a resistor with a value greater than 120 ohms if the pdf is Gaussian?

A plot of the Gaussian pdf is shown in Figure A.4. The standard normal random variable is given by

$$z = \frac{x-101}{10}$$

The value of interest is $X = 120$ ohms, which maps to $Z = 1.90$. Thus the desired probability is $Q(1.90) = 0.0287$, or 2.87% chance of selecting a resistor that is 120 ohms or greater.* □

Example A.4. Consider the problem in Example A.3, where a Gaussian pdf applies, with $\mu = 101$ and $\sigma = 10$. If the resistors are tested and any that fall

---

* Note that the probability of greater than 120 ohms or 120 ohms or greater are equivalent since the probability of selecting a resistor of exactly 120 ohms is zero.

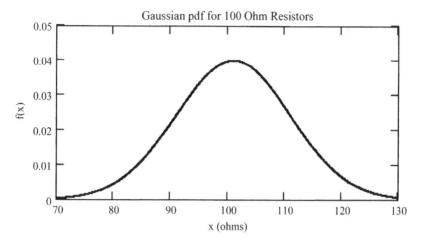

Figure A.4 Gaussian pdf for the lot of 100-ohm resistors, with mean value of 101 ohms and standard deviation of 10 ohms.

outside of the range from 90 to 110 ohms are rejected, then what percentage of the resistors are passed?

Referring to Figure A.4, the region of interest is between 90 and 110 ohms. The easiest way to determine the percentage of resistors passed is to compute the percentage rejected and subtract from 100.

For the number of resistors greater than 101 ohms, the area or probability is $Q(Z_B)$, where $Z_B = (X - 101)/10 = 0.9$. For for the number of resistors less than 90 ohms, the area or probability is best determined by using symmetry. In this case, $W = (X - 101)/10 = -1.1$. To compute the probability for region A then, one may use the value of $W = +1.1$ (the reader should satisfy himself that this is true). Thus

$$P(\text{pass}) = 1 - Q(0.9) - Q(1.1)$$

Using values from Table A.1, the result is that the probability of passing is 0.6802, or about 68% of the resistors will pass. $\square$

JOINT PROBABILITY

Given two possible events A and B in the event sample space. If the occurrence of A precludes the occurrence of B and vice versa, then A and B are said to be *mutually exclusive*. From a probability standpoint, this is written as

$$P(A + B) = P(A) + P(B)$$

The concept of the Venn diagram is a useful visualization aid. The union of two events is written as

$$A \cup B \quad \text{or} \quad A + B$$

and means the occurrence of A or B or both A and B. The intersection of two events is described as

$$A \cap B \quad \text{or} \quad AB$$

and means that both A and B occur. In general,

$$P(A + B) = P(A) + P(B) - P(AB)$$

This is illustrated by the Venn diagram in Figure A.5. If A and B are mutually exclusive, $P(AB) = 0$ since the probability of both events occurring is zero. If A and B are exhaustive (i.e., the sample space consists entirely of A and B) and mutually exclusive, then the following are true:

$$P(A) = 1 - P(B)$$
$$P(\overline{A}) = P(B)$$
$$P(\overline{B}) - P(A)$$

If A and B are statistically independent, then the occurrence of one provides no information about the occurrence of the other. It is important to understand that mutually exclusive events are not independent and vice versa!

One useful definition of independence is that the joint pdf for two independent random variables can be factored into the product of two single-variable pdf's. Thus A and B are independent, then the probability of their intersection (both events occuring) can be written as

$$P(AB) = P(A)P(B)$$

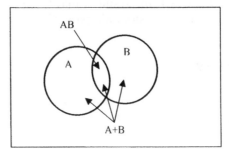

Figure A.5 Venn diagram showing events A and B and their probability of occurrence.

CONDITIONAL PROBABILITY

The probability of A given that B has already occured is

$$P(A|B) = \frac{P(AB)}{P(B)}$$

and similarly,*

$$P(B|A) = \frac{P(AB)}{P(A)}$$

By combining these two expressions, one may write

$$P(A|B)P(B) = P(B|A)P(A)$$

or

$$P(B|A) = \frac{P(A|B)P(B)}{P(A)}$$

This is Bayes' rule, which permits the determination of $P(B|A)$ in terms of $P(A|B)$, $P(A)$, and $P(B)$. This is widely used in probabilistic analysis, including detection theory.

Example A.5. Given three urns, A_1, A_2, and A_3, filled with black and white balls. The urns are selected with equal probability. The contents of the urns are as follows:

$$
\begin{array}{lll}
A_1 & 4 \text{ white} & 1 \text{ black ball} \\
A_2 & 3 \text{ white} & 2 \text{ black balls} \\
A_3 & 1 \text{ white} & 4 \text{ black balls}
\end{array}
$$

If an urn is randomly selected and a ball is drawn and seen to be black, what is the probability that the selected urn was A_3? Clearly,

$$P(A_3) = \frac{1}{3} \qquad \text{(one of three)}$$

$$P(Black|A_3) = 0.8 \qquad \text{(four of five)}$$

$$P(Black) = \frac{7}{15} \qquad \text{(seven of fifteen)}$$

---

* Since $P(AB) = P(BA)$.

Next, using Bayes' rule to find the probability that the selected urn was A_3 given that the ball was black,

$$P(A_3|Black) = \frac{P(Black|A_3)P(A_3)}{P(Black)}$$

$$P(A_3|Black) = \frac{\frac{4}{5} \cdot \frac{1}{3}}{7/15} = 4/7$$

So, once a black ball is drawn, the probability of the selected urn being A_3 goes from 1/3 to 4/7. □

The next example illustrates how Bayes' rule can be used in implementing an optimal detection scheme for a communication system.

Example A.6. A binary communication channel carries data that is comprised of logical zeros and ones. Noise in the channel may cause an incorrect symbol to be received. Assume that the probability that a transmitted zero is received as a zero is

$$P(R_0|T_0) = 0.95$$

and the probability that a transmitted one is correctly received is

$$P(R_1|T_1) = 0.9$$

The probabilities of each symbol being transmitted are

$$P(T_0) = 0.4$$
$$P(T_1) = 0.6$$

What is the probability that a one will be received, and if a one is received, what is the probability that a one was transmitted? The probability that a one will be received is

$$P(R_1) = P(R_1|T_1)P(T_1) + P(R_1|T_0)P(T_0)$$
$$P(R_1) = 0.9(0.6) + (1 - 0.95)0.4 = 0.56$$

The probability that a one was transmitted given that a one was received is found by applying Bayes' rule,

$$P(T_1|R_1) = \frac{P(R_1|T_1)P(T_1)}{P(R_1)}$$

$$P(T_1|R_1) = \frac{0.9(0.6)}{0.56} = 0.964 \quad □$$

JOINT PROBABILITY DENSITY FUNCTIONS

Joint pdf's are expressed as $f_{XY}(x, y)$. The marginal probability is the pdf in one random variable when all possible values of the other random variable are considered, that is,

$$f_X(x) = \int_{-\infty}^{\infty} f_{XY}(x, y)dy$$

Applying Bayes' rule, the joint conditional pdf's can be written as

$$f_{X|Y}(x|y) = \frac{f_{XY}(x, y)}{f_Y(y)}$$

$$f_{Y|X}(y|x) = \frac{f_{XY}(x, y)}{f_X(x)}$$

It should be apparent that $f_{XY}(x, y) = f_X(x)f_Y(y)$ if and only if x and y are independent.

Joint pdf's lay the groundwork for examining covariance and correlation, two topics of central importance in probabilistic analyses. The covariance of x and y is given by

$$\mu_{XY} = E\{(x-\bar{x})(y-\bar{y})\}$$

and provides a measure of the correlation of the two random variables x and y. The correlation coefficient is defined as

$$\rho_{XY} = \frac{\mu_{XY}}{\sigma_X \sigma_y}$$

where μ_{XY} is defined as above and σ_X and σ_Y are the standard deviations of the marginal pdf's for x and y, respectively. If x and y are independent, then they are also uncorrelated, $\rho_{XY} = 0$ and $\mu_{XY} = 0$. That is, independence implies uncorrelation. The reverse is not true in general, although for Gaussian random variables, uncorrelation does imply independence; this can be seen by looking at the expression for a jointly random Gaussian pdf, with the correlation coefficient, ρ_{XY}, set equal to zero. When the correlation coefficient is zero (meaning that x and y are uncorrelated), then the joint Gaussian pdf can be factored in to a Gaussian pdf for, x, and another one for, y, which is the definition of independence. It must be emphasized that this is not true in general for an arbitrary pdf. To see this, consider

$$f_{XY}(x, y)$$

$$= \frac{1}{2\pi\sigma_X\sigma_Y\sqrt{1-\rho^2}} \exp\left(-\frac{\left[\left(\frac{x-\mu_X}{\sigma_X}\right)^2 - 2\rho\left(\frac{x-\mu_X}{\sigma_X} \cdot \frac{y-\mu_Y}{\sigma_Y}\right) + \left(\frac{y-\mu_Y}{\sigma_Y}\right)^2\right]}{2(1-\rho^2)}\right)$$

setting $\rho = 0$ yields

$$f_{XY}(x, y) = \frac{1}{2\pi\sigma_X\sigma_Y} \exp\left(-\frac{\left[\left(\frac{x-\mu_X}{\sigma_X}\right)^2 + \left(\frac{y-\mu_Y}{\sigma_Y}\right)^2\right]}{2}\right)$$

which can then be factored into $f_X(x)f_Y(y)$.

The Rayleigh pdf describes the envelope (magnitude) of a complex Gaussian passband signal [9,10]. The Rayleigh pdf is given by

$$p(r) = \begin{cases} \dfrac{r}{\sigma^2} e^{\frac{-r^2}{2\sigma^2}}, & r \geq 0 \\ 0, & r < 0 \end{cases}$$

Unlike the Gaussian pdf, the probabilities can be determined by evaluating the integral of the pdf directly. The parameters of the Rayleigh density are expressed based on the variance of the underlying independent Gaussian random variables. The mean and variance of a Rayleigh random variable are given by

$$\mu_r = \sigma\sqrt{\frac{\pi}{2}}$$

$$\sigma_r^2 = \sigma^2\left(2 - \frac{\pi}{2}\right)$$

The Rayleigh random variable can be viewed as the magnitude of a complex signal with both the in-phase and quadrature components being Gaussian random variables with zero mean and identical variances, σ^2. The Rayleigh pdf is frequently used to describe the signal variations that are observed when there are many reflective paths to the receiver and no direct path. This is a good model for a multipath channel, based on the Central Limit Theorem since both the in phase and quadrature components of the received signal are the sum of many independent reflections.

The Ricean pdf [11, 12] is used to model a received signal that has a dominant return and multiple reflected returns [13]. The Ricean pdf is given by

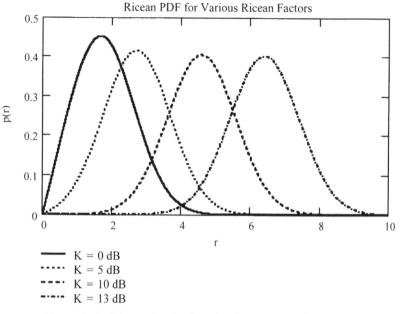

Figure A.6   Ricean density function for several values of K.

$$p(r) = \begin{cases} \dfrac{r}{\sigma^2} e^{\frac{-(r^2+A^2)}{2\sigma^2}} I_0\!\left(\dfrac{Ar}{\sigma^2}\right), & A \geq 0, \quad r \geq 0 \\ 0, & r < 0 \end{cases}$$

where

  σ^2 is the total reflection **power** that is received (i.e., the variance of the multipath signal)

  A is the **amplitude** of the direct or specular component

  I_0 is the modified Bessel function of the first kind, order zero [14]

The *Ricean factor* is defined as

$$K = 10 \log\!\left(\frac{A^2}{2\sigma^2}\right) \quad \text{dB}$$

The Ricean factor, K, is the ratio of the dominant component power $(A^2/2)$ to the multipath power, σ^2, expressed in dB. Note that A can be expressed in terms of the rms received multipath signal, σ, and the Ricean factor as follows:

$$A = \sigma\sqrt{2 \times 10^{K/10}}$$

The Ricean pdf provides a method of determining the probability of any given fade depth if any two of the three parameters, A, σ, and K are known. Figure A.6 shows plots of the Ricean probability density function for several values of K. Note that as K gets large, the Ricean pdf begins to look like a Gaussian pdf with a large mean. Of course, theoretically it can never become Gaussian because the Gaussian pdf has infinite tales and the Ricean pdf is zero for r less than zero. Nonetheless, for practical applications, once K exceeds about a factor of 10, the Gaussian pdf is a good approximation.

REFERENCES

1. W. B. Davenport and W. L. Root, *An Introduction to the Theory of Random Signals and Noise*, IEEE, New York, 1987, pp. 84–85.

2. A. Papoulis, and U. Pillai, Probability, *Random Variables and Stochastic Processes*, 4th ed., McGraw-Hill, New York, 2002, pp. 6–7.

3. P. Z. Peebles, *Probability, Random Variables and Random Signal Principles*, 4th ed., McGraw-Hill, New York, 2001, pp. 13–14.

4. W. B. Davenport and W. L. Root, *An Introduction to the Theory of Random Signals and Noise*, IEEE, New York, 1987, pp. 81–84.

5. A. Papoulis and U. Pillai, Probability, *Random Variables and Stochastic Processes*, 4th ed., McGraw-Hill, New York, 2002, pp. 278–284.

6. P. Z. Peebles, *Probability, Random Variables and Random Signal Principles*, 4th ed., McGraw-Hill, New York, 2001, pp. 125–128.

7. A. Papoulis and U. Pillai, Probability, *Random Variables and Stochastic Processes*, 4th ed., McGraw-Hill, New York, 2002, p. 216.

8. P. Z. Peebles, *Probability, Random Variables and Random Signal Principles*, 4th ed., McGraw-Hill, New York, 2001, pp. 122–125.

9. A. Papoulis and U. Pillai, Probability, *Random Variables and Stochastic Processes*, 4th ed., McGraw-Hill, New York, 2002, pp. 190–191.

10. P. Z. Peebles, *Probability, Random Variables and Random Signal Principles*, 4th ed., McGraw-Hill, New York, 2001, pp. 59–60.

11. A. Papoulis and U. Pillai, Probability, *Random Variables and Stochastic Processes*, 4th ed., McGraw-Hill, New York, 2002, pp. 191–192.

12. P. Z. Peebles, *Probability, Random Variables and Random Signal Principles*, 4th ed., McGraw-Hill, New York, 2001, pp. 399–400.

13. T. S. Rappaport, *Wireless Communications, Principles and Practice*, 2nd ed., Prentice-Hall, Upper Saddle River, NJ, 2002, pp. 212–214.

14. L. C. Andrews, *Special Functions of Mathematics for Engineers*, 2nd ed., McGraw-Hill, New York, 1992, pp. 287–290.

Introduction to RF Propagation, by John S. Seybold
Copyright © 2005 by John Wiley & Sons, Inc.

Printed and bound by CPI Group (UK) Ltd, Croydon, CR0 4YY

16/04/2025

14658581-0005